Climate Change and its Causes, Effects and Prediction

Paleoecology of Peatlands: Quaternary Climate Reconstruction from Hungary

CLIMATE CHANGE AND ITS CAUSES, EFFECTS AND PREDICTION

Additional books in this series can be found on Nova's website under the Series tab.

Additional E-books in this series can be found on Nova's website under the E-book tab.

CLIMATE CHANGE AND ITS CAUSES, EFFECTS AND PREDICTION

PALEOECOLOGY OF PEATLANDS: QUATERNARY CLIMATE RECONSTRUCTION FROM HUNGARY

GUSZTÁV JAKAB,
PÁL SÜMEGI
AND
ERZSÉBET SZURDOKI

Nova Science Publishers, Inc.
New York

Copyright © 2011 by Nova Science Publishers, Inc.

All rights reserved. No part of this book may be reproduced, stored in a retrieval system or transmitted in any form or by any means: electronic, electrostatic, magnetic, tape, mechanical photocopying, recording or otherwise without the written permission of the Publisher.

For permission to use material from this book please contact us:
Telephone 631-231-7269; Fax 631-231-8175
Web Site: http://www.novapublishers.com

NOTICE TO THE READER

The Publisher has taken reasonable care in the preparation of this book, but makes no expressed or implied warranty of any kind and assumes no responsibility for any errors or omissions. No liability is assumed for incidental or consequential damages in connection with or arising out of information contained in this book. The Publisher shall not be liable for any special, consequential, or exemplary damages resulting, in whole or in part, from the readers' use of, or reliance upon, this material. Any parts of this book based on government reports are so indicated and copyright is claimed for those parts to the extent applicable to compilations of such works.

Independent verification should be sought for any data, advice or recommendations contained in this book. In addition, no responsibility is assumed by the publisher for any injury and/or damage to persons or property arising from any methods, products, instructions, ideas or otherwise contained in this publication.

This publication is designed to provide accurate and authoritative information with regard to the subject matter covered herein. It is sold with the clear understanding that the Publisher is not engaged in rendering legal or any other professional services. If legal or any other expert assistance is required, the services of a competent person should be sought. FROM A DECLARATION OF PARTICIPANTS JOINTLY ADOPTED BY A COMMITTEE OF THE AMERICAN BAR ASSOCIATION AND A COMMITTEE OF PUBLISHERS.

Full color presentation of graphics is available in the E-book.

LIBRARY OF CONGRESS CATALOGING-IN-PUBLICATION DATA

Jakab, Gusztav.
  Paleoecology of peatlands : Quaternary climate reconstruction from Hungary / Gusztav Jakab, Pal S|megi, Erzsibet Szurdoki.
    p. cm.
  Includes index.
  ISBN 978-1-61728-220-1 (softcover)
  1. Paleoecology--Hungary. 2. Paleoecology--Quaternary. 3. Bog ecology--Hungary. 4. Peat bogs--Hungary. I. S|megi, P. II. Szurdoki, Erzsibet. III. Title.
  QE720.2.H8J35 2009

Published by Nova Science Publishers, Inc. ✝ *New York*

CONTENTS

Preface		**vii**
Chapter 1	The Quaternary Period	1
Chapter 2	The Characteristics of Plant Macrofossil Communities	5
Chapter 3	Fossil Mosses used as Proxies for Detecting Past Climatic Changes	9
Chapter 4	Material and Field and Lab Methods of Plant Macrofossil Studies	15
Chapter 5	Biogeography and Ecology of Peatbogs and Sphagnum in Hungary	19
Chapter 6	Reconstruction of Wetness Dynamics from the Sirok Nyírjes Peatbog	23
Chapter 7	Palaeoclimatic Signals and Anthropogenic Disturbances from the Nagybárkány Peatbog	27
Chapter 8	Review of the Peat-Based Paleoclimate Reconstructions from Hungary	33
Acknowledgments		41
References		43
Index		55

PREFACE

The Quaternary covers the last 2.5 million years in Earth history. This unique period is well known for a record of oscillating climatic parameters. If you wish understand the trajectory of future climatic changes triggered by human activities, one should also have a clear picture of the climate of the past as well. Fluctuating climates are reflected in peatbog profiles. Paleoecological studies using plant macrofossils, like bryophyte remains, have an important role in the reconstruction of past hydrological changes in lakes and peatbogs. Plant macrofossil analysis has been used most frequently in the oceanic regions of Europe, where the moisture gradient is reflected clearly in different *Sphagnum* taxa. The method of bog surface wetness predictions has not been adapted so far for the characterization of continental peatbogs. Hungary is located on the southern limit of *Sphagnum* dominated peatbogs. Peatbogs are restricted to the moister regions of the country. Holocene climatic events, like severe droughts, caused significant changes in mire development and as such are traceable in the paleoenvironmental record of these bogs.

Chapter 1

THE QUATERNARY PERIOD

The Quaternary is the youngest period in Earth history, which started approximately 2.5-2.6 mya and has not yet ended. The Quaternary is traditionally divided into two parts: the Pleistocene and the Holocene. The most important distinctive feature of the Pleistocene is the cyclical advancement and retreat of continental and montane ice-sheets as a result of fluctuating temperatures. As shown by the geological record of ice rafting debris and resulting moraines, in extreme cases ice masses could have covered more than 10% of the entire planet; i.e. the total mass of these ice bodies was clearly a lot greater in the past than today (>50 million km^2). Fluctuations in the climate triggered transformations not only in the area of landmasses under constant ice cover, but in the adjacent periglacial belts as well. The emergence of temperate loess areas can also be linked to these cold and dry zones adjacent to the ice bodies, which were as such characterized by a scant vegetation cover. The iterative pattern observable in the succession of loess-paleosol sequences captured the interest of Milutin Milankovitch as well, who made an attempt to correlate this fluctuating pattern with the one inferred for the climate from the analysis of past glacier valleys and inland moraine belts and horizons. The Scottish geologist James Croll (Croll 1875) was the first to postulate the idea that the observable changes in the planet's movements can actually influence its climate. Milankovitch (1930) managed to work out a mathematical model of these changes by comparing the pattern seen in the record of past glacial deposits and the parameters of the Earth's movement and preparing a theory called the ice age calendar. Insolation minima attributable to these astronomical forces was correlated with the cool periods of the ice age termed as glacial, while maximum insolation values were seen to correlate with intermittent warmer periods of interglacials. The eccentricity, axial tilt, and precession of the Earth's orbit vary in several patterns, resulting in 100,000-year ice age cycles of the Quaternary glaciation over the last few million years. The newly available geological and paleoecological record for the past 800 ky seem to have corroborated the ideas put forth in Milankovitch's work. Detailed investigations on the record of the Quaternary have also pointed to another important issue of past climatic pattern. Namely, that besides these astronomical forces, fluctuations in various other interrestrial parameters have also play a crucial role in the emergence of successive glacials and interglacials. The most important of these latter forces were those of the concentration of land masses around the poles, or the emergence of larger highlands and plateaus like that of the Tibetian Plateau, or the shape and distribution of the landmasses on the planet, which all have fundamental influences on the pattern of movement of air masses and ocean currents (Kutzbach et al. 1997).

The relationship between the pattern observable in glacier valleys at different elevations, highland and inland moraine belts as well as the succession of loess-paleosol sequences was fairly early noticed. Nevertheless, the most important evidence of past climate changes have come to light from deep-sea cores. Changes observable in the clay, organic content as well as fluctuations of isotopes of various elements built-into the shells of carbonate-shelled protists retrieved from these cores were all correlable with the pattern observable in the terrestrial records. High-resolution geochemical and isotope geochemical investigations of yearly accumulating ice laminas of inland ice masses as well as those implemented on successively deposited lacustrine layers complemented by various detailed sedimentological and paleoecological studies of the latter deposits have yielded outstanding information regarding past climatic patterns. Another important discovery was the interrelatedness of past climatic fluctuations and the pattern seen on tree rings as well as the successive deposits of glacial lakes deposited during the summer and winter periods of the year and as such characterized by different physical and chemical parameters (Lowe - Walker 1997).

Chronological studies are also indispensable to correctly establish a trajectory of past climatic changes, as these data enable us to correlate the changes observable in various parts of the planet. The most well-known absolute dating technique for the period of the Quaternary is that of radiocarbon analysis, based on the development of 14C isotopes as a result of the interaction of charged particles coming from the Sun with nitrogen located in the uppermost part of the atmosphere (Libby 1946). The resulting carbon isotope quickly turns into carbon-dioxide and is distributed relatively evenly in the atmosphere. The isotope of carbon is assimilated by plants via photosynthesis and through them it gets into other components of the biosphere as well. After the death of the organism, loss of ^{14}C due to radioactive decay is no longer counterbalanced by assimilation, and the amount of ^{14}C starts to decrease exponentially (half-time: 5730 years). Comparison of $^{14}C/^{12}C$ ratio in a sample of dead organic matter with that in the atmosphere enables determination of radiocarbon age for the past 70-80 ky. Several other absolute dating methods for the deposits of the Quaternary are also widely applied, like that of electrospin resonance, TL or U/Th isotope measurements (Lowe - Walker 1997).

Chronological studies of the Quaternary sequences enabled a correct correlation of synchronous global events, and the exact spatial and temporal delineation of environmental changes. These chronological studies shed light to the varying nature of these transformations. Namely, that some events are synchronous and relatively short-lived and represent a single time period of a sudden transformation, which is traceable to large distances, like that of volcanic eruptions and the emergence of resulting tephra horizons. Meanwhile, other changes seem to appear at different times in different parts of the planet. These latter events are referred to be time-transgressive and excellent examples are the retreat of a glacier or the pulsating spreading of environmental-dependent species as a result of the transformation of the regional environment. Thus the pattern of first appearance and last appearance of various Quaternary organisms in various parts of the planet must be used for age determination with caution (Lowe - Walker 1997).

Conversely, via the identification of collective appearance and retreat of taxa characterized by certain differing ecological needs, transformations of the environment can be more or less reliable modeled. One of the most dynamically developing fields of Quaternary research paleoecology is dealing with the investigation of such issues and problems. Paleoecological investigations enable us to draw a picture of the climate, flora and fauna, soils of the past and reveal the transformations observable in these. Complemented with the results of chronological studies, these investigations helped us reveal the environmental changes for the past 2.5 my. These studies have also shed light an important aspect of environmental changes. Namely, that these events are relatively fast geologically speaking and a complete transformation of the environment can occur within even a couple thousand years (Birks -Birks 1980).

Chapter 2

THE CHARACTERISTICS OF PLANT MACROFOSSIL COMMUNITIES

Plant macrofossils represent dead plant remains embedded in various deposits and visible by the naked eye as well. Frankly, the correct determination of this calls for the utilization of precise stereo microscopes. The most frequently studied plant macrofossils are those of seeds and fruits of flowering plants. Other important remains are those of the other vegetative parts like root and leaf fragments, stalks, buds, charcoal and fossil mosses as well. Several detailed studies are readily available on the ecology of fossil mosses (Birks 1982, Dickson 1986, Janssens 1983a, 1987, 1990, Miller 1983).

Due to their size moss remains are generally present in the fossil material as either macro- or microfossils. The small spores of mosses are generally preserved as microfossils and as such are frequently identified during the course of palynological investigations. The most frequently occurring microfossils are those of the spores of the taxa *Sphagnum* and various brown mosses (*Drepanocladus, Calliergon*). Although a well-known atlas for the identification of moss spores is readily available (Boros et al. 1993), palynological studies vey rarely attempt to identify moss taxa to the species level, due to various problems. Determinations are mainly restricted to the various taxa of *Sphagnum* mosses. Moss tissues are relatively fragile and as such are prone to rapid decay. Consequently moss remains from older periods of the Earth history preceding that of the Quaternary are relatively rare. Among fortunate circumstances, like in case of a rapid burial of organic remains there is a chance to study the cell structures as well. The observable micro features of the leaves, like those of the shape of the angular cells, or the thickness and form of cell walls, or the shape of the leaf margin or marginal cells enable us to identify remains to the level of species as well. As most of the taxa retrieved from Quaternary deposists are present in the modern flora as well, monographs of the present-day taxa are readily usable for taxonomic identification. Nevertheless, compendia for fossil taxa of mosses are also available (Janssens 1987, 1983b). Most of the remains are very rarely intact, but rather show signs of transformations and alterations. As the remains are generally dissected, there is no possibility to observe the original distribution and shape of leaves and sprouts. Very often the angular cells themselves also disappear hampering correct species-level identification. Thus the use of a modern reference collection is indispendable for the correct identification.

Quaternary moss remains are generally referred to as being in a subfossil state, namely that their cells have not yet undergone a complete geochemical alteration of fossilization characterizing other older remains. One trivial exception is moss remains preserved in freshwater carbonates (travertines). Nevertheless, this unique preservation state does not prevent the identification of these types of moss fossils (Boros 1925, 1952).

Catchment basins of lakes and mires of varying high-organic content are ideal sites for trapping moss remains. The remains themselves may be preserved in situ or could have suffered transportation before burial. The former is characteristic of mires, while the latter is that of lakes (Miller 1983).

Detailed studies of fossil mosses are available from the temperate zone of the northern hemisphere like Europe (Abramova & Abramov 1962, Boros 1952, Dickson 1973, Pilous 1968, Szafran 1952, Jasnowski 1957, Odgaard 1981, Janssens 1977, etc.) or North America (Miller 1976, Janssens 1983b). According to Miller (1983), ca. 18% of the modern taxa is also known from deposits of the Quaternary in the North American continent. Conversely, this value is 31% for the British Isles, although it must be noted that the richness of modern taxa is a lot lower in this latter case as well. Most data is avalaible on the moss flora of the Late Glacial. Conversely, relatively little is known about the older Pleistocene or younger Holocene species.

What are the most important distinctive features of Late Glacial and Holocene moss floras? One of the most striking feature is that taxa presently having an Artic or Alpine distribution also turned up on the lowlands at large distances from their modern distribution areas during the ice ages. This was followed by a rapid retreat with the improving climatic conditions in the Holocene (Herzog 1926). When these taxa, characterized by a northern distribution area are found in the modern taxa of their Late Glacial habitats, they are generally referred to as relict species. Relicts are those taxa, which were characteristic and dominant elements of a local biota in a period, when climatic conditions were utterly different at the site from the modern one, but also managed to survive to modern times due to ideal microclimatic conditions. Among mosses, the most often cited relict taxa are those of the disjunct temperate populations of Arcto-Alpine elements. A major task is to determine whether the identified species was a permanent element of the flora or a new immigrant, as mosses are relatively easy to spread thanks to their unique mode of reproduction via spores or other propagula. The species *Polytrichastrum alpinum* is a representative element of the Late Glacial flora of Denmark, which also turns up in the local modern flora as well. As the modern climatic conditions there are far from ideal for the referred taxon, its presence in the area might be interpreted as ice-age relicts at first sight. However, as known from the geological record, the present-day habitats of the taxa were inundated by the sea 5000 years ago. Thus this element must be a Late Holocene immigrant to the local flora (Odgaard 1980).

There is a group of rare moss taxa within the flora of the European continent, which is considered as ice age relicts today. These are mostly made up of Boreal-Arctic rich fen elements: *Calliergon trifarum, Cinclidium stygium, Helodium blandowii, Meesia longiseta, Meesia triquetra, Paludella squarrosa*. After Herzog (1926) these species are widely accepted as being glacial relicts (eg. Gams 1932, Szafran 1948, Stefureac 1962, Rybníček 1966, Warncke 1980). Odgaard (1988) highly questions this assumption bringing up counter evidence through the Holocene evolutionary history of the moss taxon *Meesia triquetra*. The species *Meesia triquetra* tends to appear at a given stage of wetland succession and can be only regarded as being a relict to the area under investigation if the successional development of the vegetatin lasted long enough. This taxon generally turns up coevally with the development of floating mats („schwingmoor") in lacustrine basins, or at a stage when eutrophic mires turn into oligotrophic peatlands. This latter phenomenon is often triggered by some sort of human activity, like deforestation in the areas adjacent the catchment basin (Grosse-Brauckmann et al. 1973, Rybníček - Rybníčková 1974, Rybníčkova 1974). Hall (1979) published data on the Holocene history of the taxon *Meesia longiseta*. The appearance of this latter taxon, similarly to *Meesia triquetra*, can also be connected to a certain stage in the succession of the vegetation. Namely, when the brown-moss sedge marsh turns into an acidic Sphagnum peatland. As a matter of fact neither of the two referred *Meesia* taxa can actually be regarded as ice age relict. The unique role of *Meesia longiseta* in the evolution of a plant community was identified and acknowledged via detailed macrofossil studies. *Sphagnum* peatlands tend to emerge from floating mats in the Carpathian Basin, characterized by temperately continental climatic endowments. This process is a step-wise process, where the taxon *Meesia longiseta* plays an outstanding role. During the development of the succession a transformation of the following species can be observed in the community: *Typha* → *Thelypteris palustris* → *Meesia longiseta* → *Sphagnum* spp. This entire process indicates the importance of autogenic factors resulting in the gradual emergence of oligotrophic conditions (Jakab et al. 2009, Jakab - Sümegi 2005, Jakab - Magyari 2000, Magyari et al. 2000, 2001).

Chapter 3

FOSSIL MOSSES USED AS PROXIES FOR DETECTING PAST CLIMATIC CHANGES

The detailed paleoecological investigations of fossil mosses enables us to accurately capture the prevailing conditions in some terrestrial ecosystems mainly those in littoral parts of various catchment basins. There are two major directions for the investigations and interpretations: one is restricted to the ecological needs of the individual taxa, while the other is based on the ecological requirements of ecological groups, communities in the reconstruction.

The appearance of certain taxa in the studied fossil communities may reflect special environmental conditions. The taxon *Polytrichum norvegicum* quite often appears in the Late Glacial deposits of England and Scotland, implying the presence of scattered late snowbeds in the locality (Dickson 1973, Birks – Mathewes 1978). The presence of the epiphyton *Orthotrichum obtusifolium* in the peat deposist of Two Creeks Forest Beds in WI, USA is a good indicator of poplar stands, as the bark of these trees serves as a common substrate for the representatives of this species (Miller 1976). The area of a moss taxon is generally determined by the interplay of one or some climatic variants. For example the southern lowland distribution line of the species *Polytrichastrum alpinum* is following the 16 °C July isotherm in the area; i.e. completely missing form areas characterized by hot summers (Odgaard 1980). Unfortunately, in most cases mosses are highly ambiguous indicators of the macroclimatic conditions, rather their distribution reflects the microclimatic parameters of their locality. However, transformations at the micro-scale also reflect transformations occurring at larger scales (global environmental and climatic changes). In case allochtonous communities the collective appearance of contrasting ecological character species is a rather common phenomenon. Thus a separate evaluation of allochtonous and autochtonous remains is highly desirable.

Reconstructions based on the ecological needs or parameters of entire moss communities thus yield a substantially better and reliable picture of the prevailing conditions. In this case a cluster of taxa of similar ecological parameters is created on the diagrams. such clusters are woodland or aquatic taxa, xerophylous taxa or those preferring a carbonate substrate.

This type of reconstruction is most useful in lacusrine and mire studies to capture the original vegetation. Paleobryological studies can be complemented by other paleoecological, paleobotanical studies to capture the individual stages of a hydroseries of the catchment (Birks - Birks 1980, Birks 1982, Grosse-Brauckmann 1986). Rybníček (1973) arrived at the conclusion that the composition of fossil communities is essentially the same as their modern counterparts based on a comparative study of modern and fossil marshland communities of central Europe. Being excellent syntaxonomical indicators, mosses enable us to reconstruct the entire fossil plant community when macrofossils of flowering plants are also retrieved from the deposits. The continuously growing information on the ecology of mire moss taxa is a great aid in this task (e.g. Gignac – Vitt 1990, Hájková – Hájek 2004, Hakan 1993, Janssens 1983a, b, Kooijman 1993, Slack 1994, Vitt - Chee 1990, etc.). Marshland mosses also enable us to capture fine-scale hydrological changes of the catchment, which often holds clues of transformations in the climate.

During the initial planning of sampling sites an important thing must be born in mind. The different marshlands are not equally suitable for paleoclimatological studies. One of the most essential hallmarks for the accurate assignment of coring points is the hydrology of the peatland. The hydrological parameters are naturally reflected in the morphology of the catchment of the peatland as well as the composition of the vegetation. Several peatland classification systems are known based on the hydrology, morphology or vegetation of the referred areas (Katz 1930, Mitsch - Gosselink 1993, Moore 1968, Moore - Bellamy 1974, Osvald 1925, Pakarinen 1995, Steffen 1931, Zoltai – Vitt 1995, etc.).

Hydrologically speaking peatlands can be grouped into two major groups. Ombrotrophic peatlands ('bogs') heavily rely on precipitation to establish a waterbalance, thus their evolution and distribution is primarily influenced by the climatic conditions. The water balance is mostly controlled by the mutual play of rainfall and evaporation, while the role of the vegetation and the storage capacity of the peat is negligible. Ombrotrophic peatlands are generally restricted to the Atlantic and montane areas of Europe. Water in the second group of minerotrophic peatlands ('fens') comes from the surficial watercourses as well as the groundwater besides the rainfall to the area. As such these peatlands are not as much dependant on the amount of precipitation as the previous group in maintaining a constant water balance. These two groups are further divided into various subgroups based on the morphology and the vegetation. Several classification systems are known from different parts of Europe and North America. In Atlantic Europe the ombrotrophic peatlands are often subdivided into raised bogs and blanket bogs. Raised bogs as the name implies form a positive morphology in a depression. The thinner peat layers of blanket bogs conversely follow the natural morphology of the landscape, apart from the steepest slopes naturally. In continental Europe continental peatlands are known. In the drier summer months an advancement of the arboreal vegetation is characteristic for these peatlands. Floating mats or floating fens are another typical form of continental peatlands, generally developing in the carbonate-rich or alkaline-rich shallow littoral parts of lakes and ponds (Balogh 2000a, b). There are two major types for the initiation of a peatland: in the first case a terrestrial area is turned into a peatland, which process is referred to as paludification. In the second case referred to as terrestrialization the advancement of a peatland into a lacustrine basin is observable.

The peatlands best suited for paleoclimatic reconstructions are those of ombrotrophic peatlands, which emerged via paludification. Climatic endowments, favoring the evolution of these type of peatlands are mainly restricted to the western parts of Europe under the influence of the oceanic climatic influences. As such the most peatland paleoclimatic records are known from the areas of the UK, Germany, Denmark and Sweden. Here the moisture gradient is unambiguously reflected in the distribution of certain *Sphagnum* taxa. Barber – Charman (2005) questioned the suitability of strongly continental peatlands for paleoeclimatic reconstructions.

Paleoclimatic renconstrcutions made on peatlands follow fundamentally two major approaches. One of them uses the signal of peat initiation, while the other looks for traces of compositional changes within the peat profile to make inferences about possible changes in the climate. The first approach of peat initiation signals is suitable for capturing past climatic changes at the meso (regional) and macro-scales (that of the entire continent). The initiation of peat formation at a given site is the outcome of the complex interplay of local vegetation, hydrology and the climatic endowments. Nevertheless, it must be noted that this process may very often be the outcome of some sort of human activity as well, like deforestation in the area. But at a higher scale, the initiation of peat formation is fundamentally triggered by climatic fluctuations (Halsey et al. 1998). The gist of this approach is that the age of the lowermost peat horizons is determined by looking at a large number of samples deriving from a larger area. In certain clearly identifiable periods the number of newly formed peatlands is sufficiently elevated compared to other periods, which ultimatey may indicate that these periods were characterized by cooler and wetter climatic conditions. A major drawback of this approach is that there is no evidence for the direct linear relationship between the sudden increase in the number of peatlands and the transformation of the climate to wetter and cooler conditions. Furthermore, as the majority of the original landscape is covered by peatlands anyway, new ones very rarely develop in areas of unfavorable natural endowments.

This type of approach has a long record and tradition in Canada (Campbell et al. 2000, Halsey et al. 1998, Yu et al. 2003a, b, Zoltai – Vitt 1990). According to the gained results, their the unfavorable drier climatic conditions must have prevented the expansion of peatlands up till 6000 BP. After this period an increase in the precipitation subsequently modified the local hydrological and morphological conditions enabling peatlands to reach their modern state of expansion between 3500 and 2000 BP years. Campbell et al. (2000) and Yu et al. (2003a) have managed to identify millennial and centennial-scale changes in the climate as reflected in the initiation of peat formation in Canada. These studies have clearly justified that at the scale of the continent, the development of peatlands is ultimately determined by the natural cycle of climatic changes, which is well-correlable with the findings of studies on deep-sea ocean floor and ice cores. Peatland initiation occurred in two stages in West Siberia (Smith et al. 2000, 2004). The first stage was linked to the warmings after the Late Glacial between 13,000 and 8000 BP years. The second stage took place after 5000 BP. Dry and warm conditions prevailed between these two periods. Similarly two stages of peat initiation was identified in southern Finland as well, between 800 and 7300, and 4300 and 3000 BP years. These seem to be congruent with the highstands reconstructed in the lakes of southern Sweden (Korhola 1995).

The second approach is looking for proxies reflecting transformations in the biological and chemical composition of peat sequences as signals of past climatic fluctuations. One of the most frequently used approaches in chemical analysis is that of the investigation of humification. This approach is ultimately relies on the logic that surface humidity ultimately determines the rate of decay of plant matter. When peatlands are dried out, this is reflected in a sudden increase in humic acids within the deposits. These acids are extracted from the deposits using various bases and ther concentration is determined in the solution by spectrophotometric approaches.

The most widely adopted approach in the analysis of biological components is that of the study of plant macrofossils, including mosses or those of testacea. These studies enable to identify various peatland types and past communities. However, finer-scale short-lived transformations can very rarely be linked to a given plant community. However, there is a special feature of peatland plants which can aid interpretations made on the environmental conditions. Certain species are distributed along a gradient reflecting differing water-depths. Furthermore, while certain taxa is restricted to waterlogged areas in the peatland like those of *Sphagnum cuspidatum* or *Warnstorfia fluitans*. While others show a preference to drier parts like those of *Sphagnum sec. Acutifolia* or *Pohlia nutans*. Among flowering plants representatives of *Scheuchzeria palustris* or *Menyanthes trifoliata* often turn up in the more humid periods. Conversely, representatives of the taxa *Phragmites communis, Eriophorum vaginatum, Carex elata* or peatland bushes substantially increase in drier periods. Via the utilization of various multivariate ordination methods (PCA, DCA), the less obviously identifieable moisture gradient can also be assessed in the peatland. In the next chapter a short overview of the methodology of plant macrofossil analyses is given.

Chapter 4

MATERIAL AND FIELD AND LAB METHODS OF PLANT MACROFOSSIL STUDIES

Although the importance of plant macrofossil studies in paleoecological works was relatively early identified and emphasized (Jessen - Milthers 1928, Jessen 1949), up till the 1970s macrofossil diagrams only appear as a complement of pollen diagrams. Aaby's (1976) impressive study gave an impetus to methodological improvements to be made, which initiated in the 1980s. The most important methodological works are known from the hands of Grosse-Brauckmann (1972, 1986), Birks (1980), Birks - Birks (1980), Mauquoy – van Geel (2007), Janssens (1983a, 1987, 1990), Rybníček (1973), Wasylikowa (1996), Gaillard – Birks (2007) and Beaudoin (2007). A major step ahead was the introduction of the so-called QLCMA (semi-quantitative quadrat and leaf-count macrofossil analysis technique) developed in Southampton (Barber et al. 1994), which enabled researchers to achieve an accuracy and resolution known from palynological studies in their work. Today macrofossil studies are indispensable components of Quaternary paleoecological investigations (Birks - Birks 2000, Birks 2007).

Sampling is made by using a Russian-head corer or Livingstone piston corer, generally used in Quaternary environmental historical studies (Aaby-Digerfeldt 1986, Cushing - Wright 1965, Jowsey 1966). These sampling methods yield us undisturbed cores. After transportation to the laboratory, the cores are cut lengthwise for various analyses; the sections for palaeobotanical and geochemical analyses are stored at 4 °C in accordance with international standards. Subsamples are taken in the lab involving a volume of 1-4 cm^3 in general. However, this volume may be larger as well as much as 8-10 cm^3, when we are working with samples relatively poor in organic matter. Sampling intervals are determined by the aims of the analysis and the inferred rate of deposition. In case of plant macrofossil studies this is generally 4 cm, but may be at the scale of mms as well. Samples are filtered using a sieve of 250 µm mesh. The more consolidated sediments are treated with 10 % KOH or NaOH for 5-10 minutes.

In our work a modified version of the QLCMA technique was adopted (Barber et al. 1994, Jakab et al. 2004). Organic remains from peat and lacustrine sediments rich in organic matter can be divided into two major groups. Some remains can be identified with lower ranking taxa (specific peat components), while others cannot be identified using this approach (non-specific peat components). Sediment samples can contain significant amounts of non-specific peat components, which reveal much about the hydrologic conditions and chemical composition of the area in which the sediment accumulated. The most important non-specific peat components are the following: undifferentiated monocotyledon remains (Monocot. undiff.), unidentified organic material (UOM), unidentified bryophyte fragments (UBF), unidetified leaf fragments (ULF), charcoal, wood. In the case of specific peat components, the remains can often be identified to the species level. They are important for reconstructing the sediment depositionary environment. The local vegetation often allows identification at the association level. The most important specific peat components are seeds, fruits, sporogons, mosses, rhizomes and epidermis (e.g. Carex species), leaf epidermis, other tissues and organs (hairs, tracheids, etc.), insect remains and Ostracod shells. The identification of herbaceous plant tissues was based on the procedure described by Jakab and Sümegi (2004b).

Concentration levels were determined by adding a known amount of indicator grains (0.5 g poppy seed, ca. 960 pieces) and by counting the poppy seeds and the remains using a stereo microscope in ten 10 mm by 10 mm quadrates in a Petri dish. Similarly to mosses, rhizomes can only be identified with a light microscope. We removed a hundred monocotyledon remains and mounted them in water on microscopic slides for determining the percentages of individual taxa and of Monocot. undiff. The values for different moss species and UBF were determined using a similar procedure. The concentration can be described with the following equation:

$$\text{macrofossil concentration} = \frac{\text{counted macrofossil (average)} \times 960 \text{ (total poppy seeds)}}{\text{counted poppy seeds (average)} \times \text{sample volume (cm}^3)} \quad (1)$$

Dominance values are depicted graphically with depth displaying information on radiocarbon age, the name of local zones and symbols of sedimentary features in accordance with the system of Troels-Smith (Troels-Smith 1955), internationally accepted for the description of unconsolidated deposits. Software packages of Psimpoll (Bennett 1992) and Syn-Tax (Podani 1993) were used for plotting the analytical results.

The excat determination of the age of the deposits is indispensable for correct paleoenvironmental reconstructions given that more than 50% of the dry peat is composed of elemental carbon, radiocarbon dating is the best suited for such purposes. Conventional radiocarbon dating has been a frequently applied technique. The gained raw BP years are calibrated to cal BP or calendar years (AD/BC). The new accelarator mass spectrometry (AMS) approach yield more precise ages and require smaller input samples too. As such smaller plant remains like that of *Sphagnum* leaves might be sufficient.

Chapter 5

BIOGEOGRAPHY AND ECOLOGY OF PEATBOGS AND SPHAGNUM IN HUNGARY

Hungary is nestled in the heart of the Carpathian Basin, characterized by moderately continental climatic conditions, enjoying the modificatory effect of various climatic influences like those of the Oceanic influences in the west and those of the Mediterranean influences in the south. The basin morphology ultimately determines the distribution of the climatic pattern. Namely, there is a gradual increase in continentality accompanied by a gradual decrease in the rainfall from the margins towards the center. The montane climatic influences prevailing in the surrounding hills of the Alps in the west and Carpathians in the north and east ultimately determine the climatic endowments of the marginal part of the basin. The most striking proxy expressing the influences of this basin morphology on the climate is the annual rate of precipitation. This value is often below 500 mm in the central driest parts of the Alföld (Great Hungarian Plains-from now on as GHP) and displays a gradual increasing trend to the north and west. The western parts of the country enjoys the highest rates of annual precipitation sometimes exceeding the value of even 900 mm. The areas of the GHP are characterized by an average annual rainfall of 500-550 mm, while those of the mid-mountains have a rate of 600-800 mm per annum (Bacsó 1959).

The general climatic endowments of Hungary are far from ideal for the emergence of *Sphagnum* peatlands. The number of localities harboring *Sphagnum* species hardly reaches the value of 50 in the country, and not even scattered occurrences of these are known from the central, driest parts of the GHP. The actual number of *Sphagnum* peatlands is below 20, the majority is being tiny with an extenson of a couple of ha. Raised bogs are completely missing. The majority of *Sphagnum* peatlands are restricted to the northern areas of the Northern Mid-Mountains and the northern GHP, as well as the estern parts of the country enjoying more precipitation thanks to the blessed work of the oceanic and montane climatic influences (Boros 1968, Szurdoki – Nagy 2002). The southernmost distributions of lowland *Sphagnum* peatlands are found in the area of the northern GHP in the entire continent. After the artificial dessication of *Sphagnum* peatlands, there is an advancement of reed, sedge and birch to these areas in accordance with the local endowments (Borhidi – Sánta 1999, Lájer 1998b). Despite all of this Hungary used to be relatively rich in peatlands and marshy areas. The extension of past marshlands exceeded 90000 hectars in the past preceding various river regulation and drainage measures. The estimated volume of peat reserves is 973 million tons (Dömsödi 1988). However, these peats are not acidic *Sphagnum* peat, but rather basic reed or sedge peat. These marshlands used to cover extensive areas of the country in the neotectonic depressions of the GHP, abandoned river channels of river Danube or in the littoral parts of larger lakes like Lake Balaton or Fertő (Neusiedler See) (Fig 1.).

Peat mosses and their habitats are very rare in Hungary and since the 19th century they have attracted a special attention. The genus *Sphagnum* is protected by law in Hungary since 1986 and most of their habitats are located in protected areas (national parks, landscape conservation areas, nature reserves or nature conservation areas).

The large amount of Hungarian peat mosses can be found in fens and willow and alder swamps where they create mixed, extent carpets, but the microtopography, as in e.g. Nordic mires (hummock, hollow, pool etc), are more or less missing. The most frequent species in bigger mires are *Sphagnum angustifolium, S. fallax, S. palustre* and *S. fimbriatum*, which are not able to build hummock-hollow system. The species (e.g. *S. capillifolium, S, rubellum, S. magellanicum*) which could create compact hummock are very rare and they have only scattered distribution in these mires, as well as hollow and pool forming species are absent except some *S. cuspidatum* occurrence. Because of rarity of *Sphagna*, all places where they live, were documented since 1900's (e.g. Boros 1924, 1937, Simon 1953, Visnya 1939, Zólyomi 1931, 1939). We know many patches of peat mosses from 0.1 to few square meters in meadows, in other wet places and also in wet, acidic forests' floor (e.g. Lájer 1998a,b, Ódor et al. 1996, Szurdoki 2003, Szurdoki - Nagy 2002, Szurdoki et al. 2000). These small patches are can be found mostly in westernmost part of Hungary (Őrség and Vendvidék, Ódor et al. 1996), where the climate is colder and wetter, because of influence of Alps, and also many in northern-eastern part which influenced by the Carpathians (Szurdoki – Nagy, 2002, Szurdoki et al., 2000).

Most *Sphagnum* occurrences are can be found in hilly area, except the south-western and the easternmost populations. *Sphagnum*-bogs are situated in western and northern part of the country, where most species could alive the unfavourable conditions of last decades. Between these two areas, the earlier viable fens and willow swamps are dried and most peat mosses are disappeared. Also, the alder-swamps of south-western region are dried (because of the drainage), and the amount of peat mosses are decreased, and they disappeared from many sites. However nowadays 24 *Sphagnum* species live in Hungary, most of them are rare or very rare, can be found only in few places. *S. auriculatum*, *S. riparium*, *S. teres* and *S. subnites* have only 1-2 occurrences and *S. rubellum*, *S. russowii*, *S. inundatum*, *S. cuspidatum* and *S. compactum* can be found in 3-5 places.

The most frequent species are *S. angustifolium*, *S. fallax*, *S. fimbriatum*, *S. palustre* and *S. squarrosum*. They create the moss carpet in mires and usually in other peat moss occurrences (smaller patches is meadows, in wet places) also, where rare species could find the suitable habitats. *S. angustifolium*, *S. fallax* and *S. palustre* live mainly in fens, but they occur in meadows along springs and streams. Many patch like occurrences of them earlier were extent fen along springs and streams, but the drainage and the dry climate of last decades degraded these habitats, and only the remnants of most frequent species could alive.

S. fimbriatum and *S. squarrosum* are the main species of willow and alder swamps and also the outer willow belt of larger fens. However, established and disappeared populations of all peat mosses were detected in last decades, only *S. fimbriatum* showed the clear expansion. Szurdoki and Ódor (2004) studied the Hungarian distribution of *S. fimbriatum*, comparing floristic data before and after 1990. Before 1990 *S. fimbriatum* was known from 13 localites, and now occurs in 32. The habitat preference of the species hardly changed during its expansion; it occurs mainly in *Salix* mires and also in tall sedge vegetation and in wet places. In most, recently found localities *S. fimbriatum* establishments have a pioneer character; it colonised earlier known (and well-studied) sites, or occupied new, young habitats. Due to the characteristic field morphology and very intensive earlier bryological studies it may be safely assumed that *S. fimbriatum* has expanded its distribution in Hungary.

Szurdoki (2005) investigated the abiotic conditions of some most frequent *Sphagnum*, in five Hungarian mires. Conductivity, pH, height above water table, Na, K, Ca and Mg concentrations were detected. The investigated mires are similar, but there were many significant differences between them in case of analytical variables, but there were only week differences inside mires. On the basis of water table, pH, and conductivity can be separated the investigated species. *S. fallax* and *S. angustifolium* are not differing from each other, which not surprise, because they live together, in mixed carpets in most investigated mires. They mainly occur in wet and acidic places with poor mineral content. *S. palustre* lives in driest places and *S. fimbriatum* in wet and less acidic places which characterized with highest mineral content.

Chapter 6

RECONSTRUCTION OF WETNESS DYNAMICS FROM THE SIROK NYÍRJES PEATBOG

The site Nyírjes Peatbog of Sirok is found in the northern part of the country at the eastern foothills of the Mátra Mts at an elevation of 250 m (Fig 1.). It covers a small area of 9000 m^2. No surficial watercourses feeding or draining the peatland is known. The basin of the peatland is fringed by a woodland of hornbeam and oak. The following plant communities are present moving from the margins towards the center: *Scirpo-Phragmitetum, Salicetum cinereae-Sphagnetum, Carici lasiocarpaea-Sphagnetum*. This peatland harbors the following peat moss taxa: *Sphagnum palustre, S. subsecundum, S. magellanicum, S. recurvum* s. l., *S. fimbriatum, S. squarrosum, S. obtusum* and *S. angustifolium*. The most common are those of *Sphagnum recurvum* s. l. and *S. palustre* (Máthé – Kovács 1958, Szurdoki – Nagy 2002). A detailed palynological work on the peatland was done by Gardner (2002).

Samples taken between the depths of 401 and 6 cm were subjected to plant macrofossil analyses. Fig 2. depicts the most important tissue and moss remains of the dominant taxa. In order to reveal the ecological-hydrological gradient of the individual macrofossil zones, a data matrix of 16 most important peat components was subjected to multivariate statistical analysis. The method of PCA was adopted following Podani (1993) using the software package SYN-TAX 5.0. The received PCA values were depicted with depth on Figure 2.

Figure 1. Peatlands (dark areas), *Sphagnum* bogs (circles) and *Sphagnum* occurrences (triangles) in Hungary and the position of the peatbogs near Sirok (1) and Nagybárkány (2).

On the basis of the gained results the following evolutionary history of the peatland can be drawn. The first emergence of aquatic conditions in the depression can be dated to 9500 cal. BP, resulting in the emergence of a relatively deep, oligotrophic lake with scant aquatic vegetation. As shown by findings of palynological studies, the lake basin was fringed by an open parkland type woodland with the dominance of *Picea, Quercus* and *Corylus* till about 8950 cal. BP. This was transformed into a woodland dominated by *Tilia* till 8300 cal. BP which then finely was transformed into a deciuous woodland dominated by *Quercus, Tilia* and *Ulmus* till 6900 cal. BP with substantial stands of *Corylus*. Despite the clearly observable transformation of the surrounding vegetation water levels were relatively stable in the basin, apart from minor fluctuations till 7500 cal. BP. A drop in the water level and peat initiation took place almost 1000 years after the development of a closed deciduous woodland. So no wonder that there is no direct link between the transformation of the vegetation of the peatland itself and the surrounding terrestrial areas. There is a gradual decrease in the water levels from 7500 cal. BP reaching an all time minimum at 6400 cal BP. Open water areas almost completely disappeared, giving way to the expansion of oak shoots to the major part of the basin. The deepest areas turned into an eutrophic marshland

and as such we must assume a gradual increase in the water level from 5800 cal. BP years yielding a tussock vegetation. During this period the peatland was fringed by a woodland of *Corylus, Quercus* and *Carpinus*.

There is another rise in the waterlevel from 5200 cal. BP resulting in the expansion of the peatland. This was accompanied by the appearance of floating mats in the expanding shallow eutrophic pond harboring peat mosses in larger substances. There is a rapid spread of *Carpinus betulus* in the adjacent closed oak woodlands at the time. Peak distributions *Fagus sylvatica* and *Carpinus* are found between 3700 and 1750 cal. BP. A similar expansion of *Sphagna* was inferred from 3900 cal. BP onwards in the basin, with the first appearance of real acidophyllic *Sphagnum* peatlands dated between 2300 and 1500 cal. BP years.

From 1500 cal. BP there is an alternating succession of *Sphagnum* peatlands with reed and sedge peatland horizons reflecting the alternations of cooler and warmer periods till the present day. Ideal *Sphagnum* peatland conditions were inferred at 1550 cal. AD with such taxa as *Sphagnum cuspidatum*. As shown by the results of Gardner (2002) there is an icrease in human influences in the area from 1750 cal. BP as seen in the drop in the amount of *Fagus* and *Carpinus* accompanied by an advent of *Quercus*. The past century was also a period of *Sphagnum* peatland expansion. The presence of clayey horizons embedding mollusk shells and carbonate concretions intercalating the peat horizons are the clear signs of soil erosion in the adjacent areas triggered by deforestation of the nearby slopes. As an outcome of these activities the amount of rainfall reaching the surface substantially increased, resulting in an increase of the water level in the bed of the peatland triggering the expansion of *Sphagna*. A similar phenomenon was described from several other European sites (Grosse-Brauckmann et al. 1973, Rybníček - Rybníčková 1974, Rybníčkova 1974).

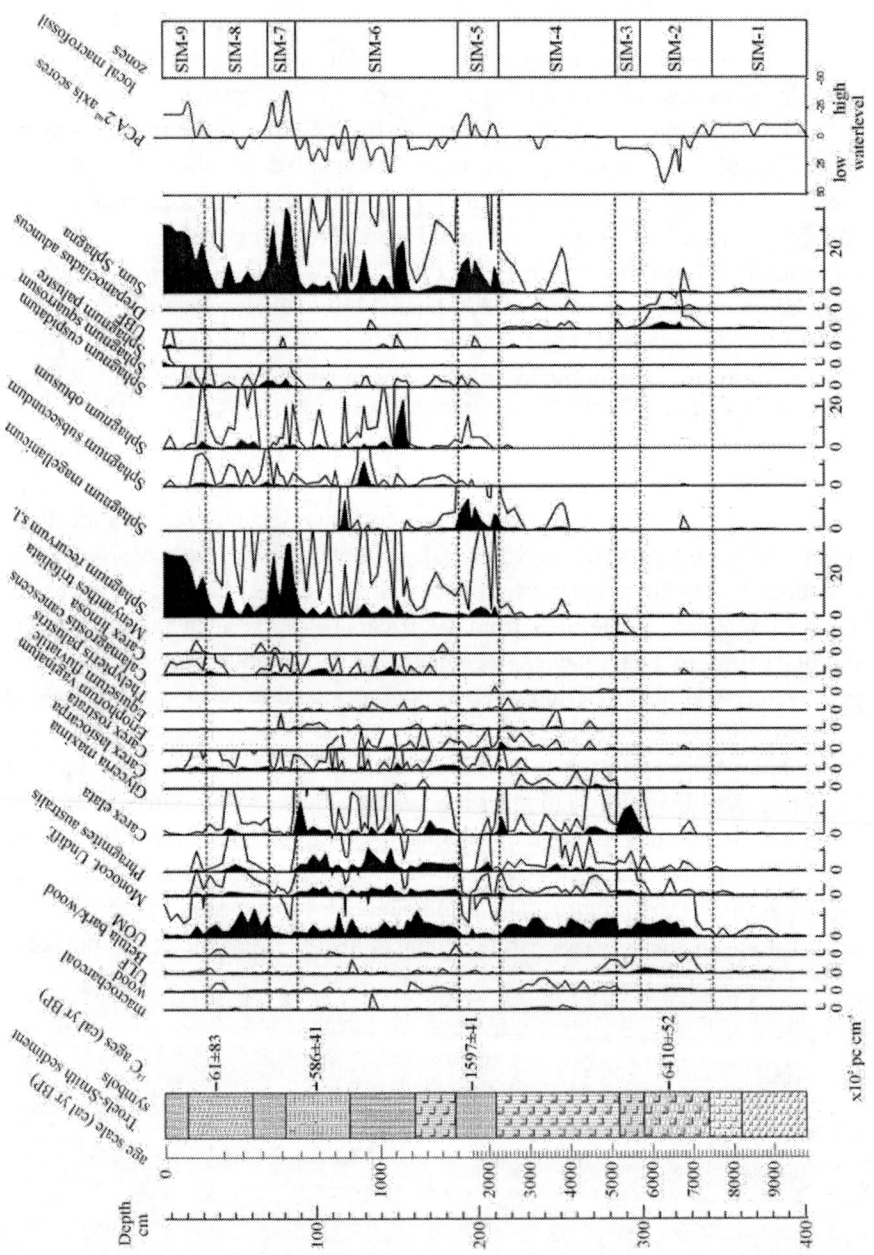

Redrawn from Jakab – Sümegi 2010.

Figure 2. Selected macrofossil diagram of Sirok Nyírjes Peatbog.

Chapter 7

PALAEOCLIMATIC SIGNALS AND ANTHROPOGENIC DISTURBANCES FROM THE NAGYBÁRKÁNY PEATBOG

The Nádas-tó (360 m a.s.l.) at Nagybárkány lies on the northern side of Mt. Hármas-Határhegy, rising to a height of 516 m in the Eastern or Pásztó Cserhát Mountains. The lakebed has an elongated, northwest orientation, with a very narrow extension in the south. Its length is roughly 100 m, its greatest width is 40 m and it covers an area of roughly 2000 m^2. Accumulation in the catchment basin of the Nádas-tó started in the Late Glacial when a mass movement (rotational landslide) occurred on the slope of the Miocene sandy and silty sediment covered land surface. A slump hollow formed in the source area between the landslide toe and the scarp which was filled up by water forming a small round-form lake. The annual rainfall is between 600-700 mm. The peatbog is solely fed by rainwater. There is not any visible watercourse in the drainage area.

The lakebed is fringed by an oak forest. Three plant communities can be distinguished in the recent bog. The central part of the bog is covered with willow swamp (Salici cinereae-Sphagnetum recurvi). This community has a low species diversity, and is characterised by the dominance of *Salix cinerea* and a carpet of *Sphagnum squarrosum*. The willow swamp is fringed by reed beds (Scirpo-Phragmitetum), except on the western shore. Tall sedge communities (Caricetum ripariae) line the reed beds. These communities are dominated by *Carex riparia*. The peatbog underwent significant changes since the preparation of the first vegetation map in 1959 (Máthé - Kovács 1959).

The willow swamp encroached on the one-time reed beds, while the reed beds replaced the former burr reed associations.

In the undisturbed sedimentary sequences peat was present to a depth of 110 cm, with an underlying water pocket down to 130 cm. Between 130–300 cm, we found peat and peat-mud with varying organic content. Between 300–340 cm, there was a silty lacustrine sediment. Fig 3. presents the most important tissue and moss remains of the dominant taxa.

The first phase (15260 – 6000 cal ys BP) of the bog's development, lasting until the middle Holocene, was determined by relatively stable trophic conditions and smaller fluctuations in the water level. The water level dropped somewhat at ca. 8800 cal yr BP and the peatbog developed, probably as a consequence of the onset of a warmer and drier climate. Based on pollen profiles, this phase coincided with the appearance of thermophilous species-rich oak forests. The water level again rose at 8000 cal yr BP and fell after 7000 cal yr BP, causing the expansion of reed beds.

The second phase of peatland development (2300 – 700 cal yr BP) started with a short but remarkable anthropogenic disturbance. Abrupt changes indicate a hiatus in the catchment basin. It appears that the unconformity was the result of human impact. Results of the radiocarbon measurements suggest a sedimentary hiatus is apparent between 240 – 220 cm. This hiatus associated with a thin layer of macrocharcoal. This and a subsequent change in lacustrine stratigraphy from mesotrophic lake to reed peat shows ca. 4000 years difference in age between adjacent samples. According to radiocarbon and sedimentological data, this event coincides with a peat-cutting in the Imperial Age when Barbarian groups (Dacs, Celts, and German tribes) lived around the lake basin (Vaday 2005). Probably, one of these ancient tribes cleared the area around the pond around 2300 cal yr BP. After this clearing, a mass of water plants covered the artificially transformed pond surface. This second phase lasted from ca. 300 cal BC until cal AD 1400. The water level decreased and the lake was encircled by a marshland belt. Following the anthropogenic disturbance this period was characterized by fluctuations in the trophic conditions, which can most likely be linked to climatic causes. Figure 4 shows the changes in the values of six peat components, which were typical for this phase, together with the historically recorded climatic phases.

Palaeoclimatic Signals and Anthropogenic Disturbances... 29

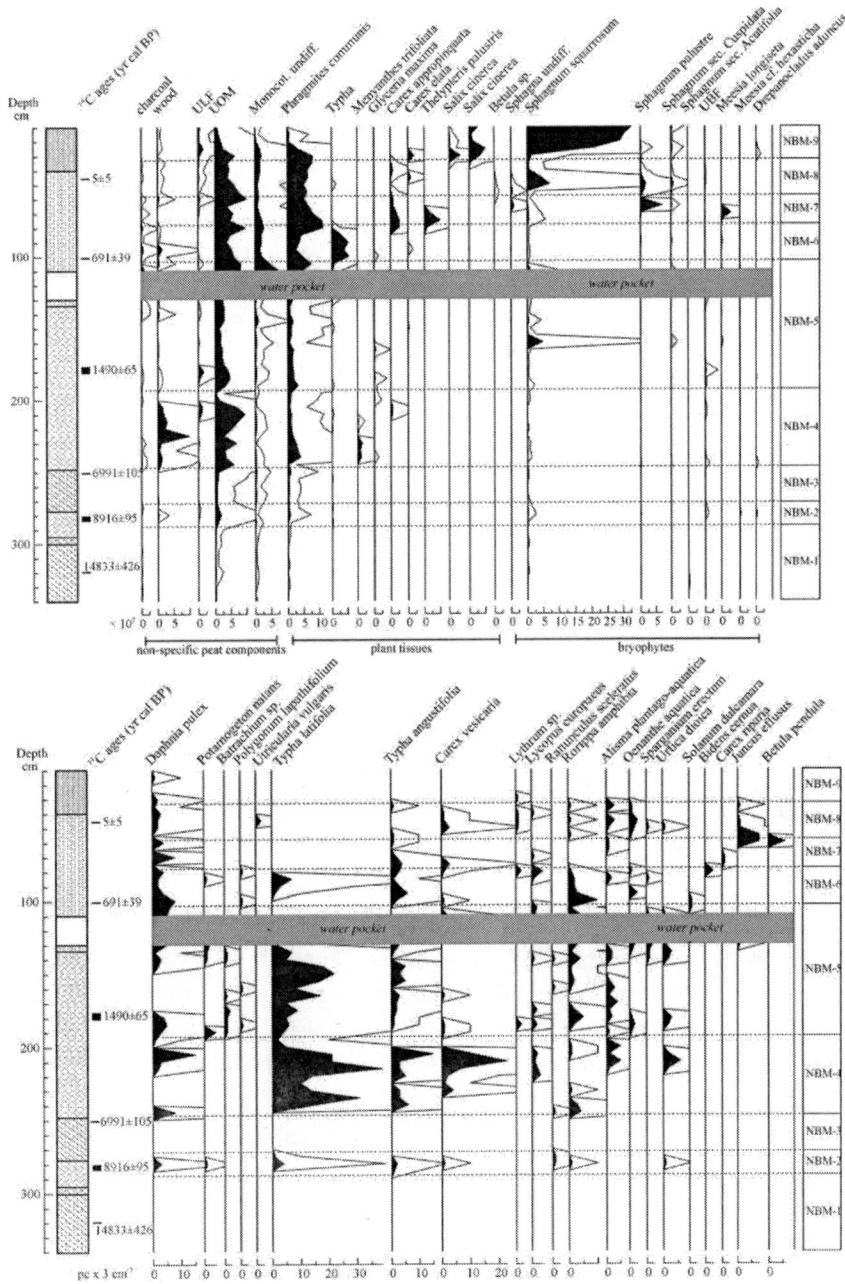

Redrawn from Jakab et al. 2009.

Figure 3. Fossil plant tissues, mosses and seeds from Nagybárkány–Nádas-tó.

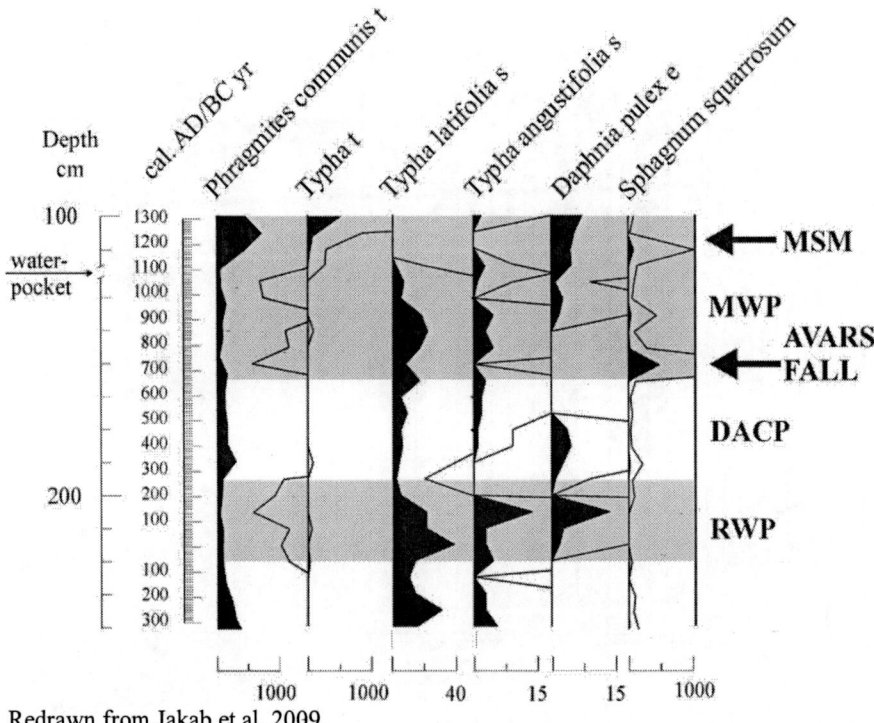

Redrawn from Jakab et al. 2009.

Figure 4. Trophic fluctuations at Nagybárkány–Nádas-tó between 330 BC and 1300 AD. The grey areas indicate periods with a warmer climate (MSM: medieval solar activity maximum, MWP: medieval warm period, DACP: Dark Ages cold period, RWP: Roman Age warm period; the interbedded water layer between 130–110 cm has been omitted, t: tissue, s: seed, e: ephippia).

Figure 4 reveals that the *Phragmites* concentrations in the sediment decreased during warmer periods in the Imperial Age and the Middle Ages, parallel with an increase of *Typha* seeds and *Daphnia* ephippia, which suggest increasing eutrophication (Keller et al., 2002). This reflects a competitive situation, characterized by alternating dominances of reed and bulrush in the lakebed. Reed and bulrush are both competitive species under favorable conditions. Bulrush is a light-demanding plant. Reeds grow higher, overshadowing bulrush and easily out competes the latter. At the same time, reeds requires relatively coarse sediment in order to anchor its roots, otherwise it collapses. Conversely, bulrush settles more easily in finer-grained sediments (Haslam 1972, Toivonen - Bäck 1989, Nurminen 2003). In the pollen record this phase is characterized by the increase of herbaceous elements which suggests the opening up of the forest canopy. It would appear that during periods of greater solar activity, the lake received more light, in part owing to the retreat of species forming higher and more closed forest canopy, like *Fagus* or *Carpinus*. The lake is fringed by high, steep slopes in the south and southeast, from where the high trees cast a shadow over the greater part of the lake. The expansion of phytoplankton at the time of greater solar activity is indicated by an increase of *Daphnia* feeding on them. The expansion of phytoplankton leads to the development of looser sediments, encouraging the spread of *Typha*.

The beginning and the close of the Medieval Warm Period saw the maximum of solar activity (Bradley et al. 2003). The first maximum, dated to AD 700–800, caused significant droughts in the Carpathian basin, and contributed to the decline and fall of the Avar Empire (Györffy 1995, Györffy & Zólyomi 1995). The end of this warm period ca. AD 1250 was marked by the so-called medieval solar activity maximum, which caused serious droughts in Europe and North America. The sudden expansion of *Sphagnum squarrosum* coincides with the time of the two maximums. *Phragmites* and *Typha* both declined around AD 800, suggesting the brief desiccation of the bed, when peat moss temporarily covered the entire lakebed. The water level decreased for a longer period of time around AD 1200–1300, enabling the expansion of the reed bed over the lake's entire surface and causing the reduction of open water.

The development of the present-day bog and the commencement of peat accumulation can be dated to the end of the Árpádian Age. Around AD 1400, the number of *Typha* rhizomes increases significantly, indicating a rise in the water level and the formation of a floating mat. The hydroseries of the bog was from this point on characterized by autogenic processes, with a tendency towards a gradual oligotrophication, triggered by climatic deterioration. Based on detailed phytogeographical studies, floating mats of *Phragmites* and *Typha* are frequent components of lake shore vegetation in Hungary. The first signs of peatbog formation from these floating mats can be seen in a massive expansion of *Thelypteris palustris*. As time goes by, peat mosses also turn up on these mats (Borhidi - Balogh 1970, Balogh 2000a, 2000b). Based on the palaeobotanical investigations, this bog development passed through phases characterized by *Typha* → *Thelypteris palustris* → *Meesia longiseta* → *Sphagnum* spp. This development shares numerous similarities with the formation of two other *Sphagnum* bogs at Csaroda–Báb-tava (Jakab - Magyari 2000) and Kelemér–Nagy-Mohos (Magyari et al. 2000, 2001), suggesting some sort of regularity in the formation of Hungarian *Sphagnum* bogs. *Meesia longiseta* preceding the expansion of Sphagna is highly interesting. According to Hall (1979), the appearance of *Meesia longiseta* can be linked to a distinct phase of wetland succession, characterized by the transformation of the brown moss-sedge floating mat into acidic *Sphagnum* bog. This species was observed during the secondary succession of abandoned lakes used for retting hemp as well.

Representatives of *Sphagna* seemed to have appeared in similar quantities during the middle part of the 17th century as today, followed by a complete drop till AD 1950. Surprisingly, the taxon *Sphagnum palustre* indicating mesotrophic conditions was present in the largest numbers. This expansion of *Sphagnum* coincides with the coldest period of the Little Ice Age Hungary (Rácz 2001), which was also the coldest time of the past 2000 years. The later expansion of *Sphagnum squarrosum* indicates the emergence of an acidophilic, eutrophic peatland in the area. *Sphagnum squarrosum* is relatively tolerant to high Ca, bicarbonate levels, and pH (Clymo 1973), and in a mineral-rich habitat with a high nutrient supply, grows very fast (Kooijman 1993).

Chapter 8

REVIEW OF THE PEAT-BASED PALEOCLIMATE RECONSTRUCTIONS FROM HUNGARY

The first Holocene millenial-scale climatic schema based on the studies of peat stratigraphy in Scandinavia a century ago (Blytt 1876, Sernander 1908). This widely applied schema was recognised as too simplistic and not realistic. Based on the data from high-latitude ice-cores Holocene climate is considered relativily benign and invariable (Oldfield 2005). Climate fluctuations of the Holocene is now a major research focus in Quaternary paleoecology, because these proxy-climate data archives may be compared with the recent global warming.

The proxy-climate records of bog restricted mostly on the western parts of Europe (Barber 2007, Barber – Charman 2005), but no discussion is made on SE Europe, including Hungary. This area appears blank on the lake data source maps pointing to the paucity of available Holocene mire surface wetness records in this region. On Figure 5 several recently obtained lake-level and mire surface wetness records are gathered from the Carpathian Basin and Central Europe.

The lakes and mires in the hilly and mountainous areas of Hungary responded to climate changes much more sensitively than those in the lowland areas. Some major climate shifts can only be detected at the lowland sites. This phenomenon can be easily explained with the water supply of the studied lakes and mires. The sites in the hilly and mountainous areas received their water supply from the precipitation and from the surficial watercourses

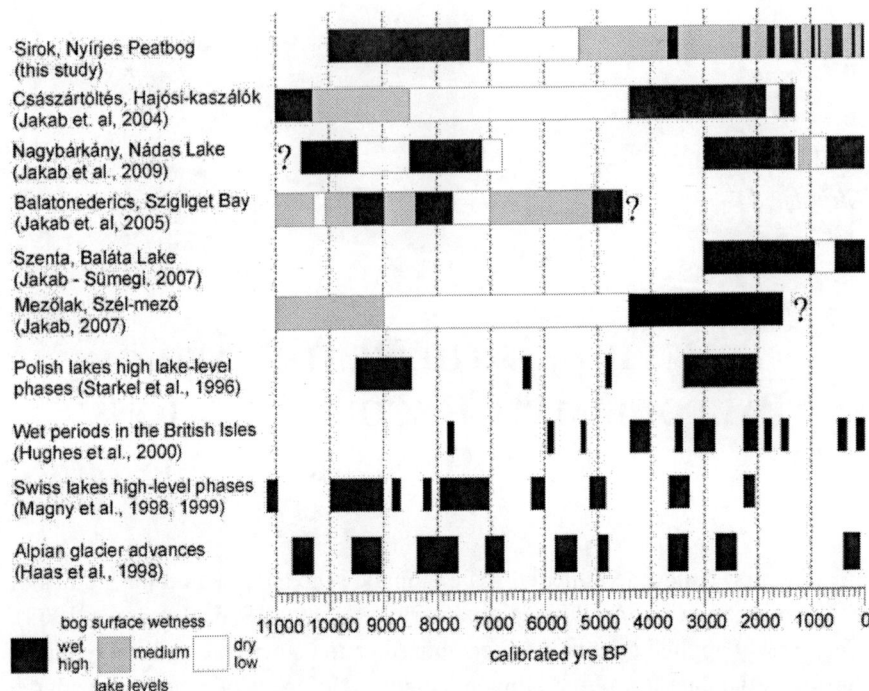

Figure 5. Comparison of Holocene bog surface wetness records with other bog surface wetness and lake-level records from Hungary and Central Europe.

primarily. Therefore they responded to climate changes more sensitively. In these cases more hydrological events with a shorter duration could have been inferred. According to our investigations, the low water levels at the Early to Mid Holocene (10200-5700 cal. BP) and the high water levels of the Neoglaciation (after 5700 cal. BP) were clearly discernible in many cases. Short humid or cold periods during the Holocene climatic optimum were also identifiable (eg.: at 8000 cal. BP), just like the low water levels of the Medieval Warm Period in the studied peat sequence.

Investigations of the Sirok Nyírjes Peatbog provides a full Holocene record of vegetation development effected by climatic changes. The emergence of an oligotrophic lake in the area was dated to 9500 cal. BP with deeper water conditions. Changes in the surficial moisture gradient of peatlands in the Carpathian Basin and those of lake level fluctuations are rather contradictory for this period. High lake level phases are known at 8500 cal. BP for the Szigliget Bay of Lake Balaton (Jakab et. al. 2005), and Lake

Nádas at Nagybárkány (Cserhát) (Jakab et al. 2009). The inferred water levels of Lake Sf Ana in Romania show a highstand at 9500 cal. BP with the emergence of a lowstand at 9000 cal. BP (Magyari et. al. 2006, 2009). Conversely, studies implemented at various sites of the GHP (Császártöltés, Mezőlak) reconstructed a long-lasting dry and warm period till about 4400 cal. BP (Jakab et al. 2004a). There seems to be substantial regional differences in the Early and Middle Holocene climate of the Carpathian Basin. These data, on the other hand, do not support Magny et al.'s hypothesis that all of mid-latitude Europe became wetter during this phase as a result of a displacement and intensification of westerly circulation (Magny et al., 2003). It is conceivable that in parts of Central-Eastern Europe the 8.2 ka event was embedded in a longer generally drier phase, within which the period between 8,200 and 8,000 years had relatively increased moisture availability, but none of the aforementioned records have yet provided evidence to support this theory. (Magyari et al. 2009).

Decreasing water levels inferred at 7500 cal. BP culminated in the driest phase of the peatland recorded at 6400 cal. BP. This period is the time of Holocene climatic optimum, when there is a substantial retreat of the Swiss Alp glaciers between 7450 and 6650 cal. BP and 6200 and 5650 cal. BP years (Joerin et al. 2008). Conversely, there is an inferred increase in the water level of Lake Sf Ana in Romania from 7500 cal. BP onwards, interrupted by a short decrease between 5500 and 5300 cal. BP (Magyari et al. 2006, 2009). Similar decreases in mean annual precipitation and the moisture availability are suggested for this period by the pollen-based climate reconstruction of Feurdean et al. (2008) and Schintchen et al. (2006), that is followed by a steep increase in mean annual precipitation from 5,300 years BP (Feurdean et al., 2008). According to Cheddadi et al. (1997) and Davis et al. (2003) the traditionally postulated Holocene climatic optimum is identifiable only in Northern Europe. This time Southern Europe was characterized by colder conditions, with Central Europe occupying a transitionary phase. This assumption is refuted by the findings of paleoecological studies made on lake and marshland basins in the Carpathian Basin.

After the climatic optimum there are two periods when substantial increase in the surface moisture gradient was observable in the referred study site: at 5800 and 5400 cal BP years. This change is congruent with the pattern observable in other lacustrine and marshland basins of the Carpathian Basin also displaying an increase in the water level. There is a sudden increase in the water level of the Lake Sf Ana from 5500 cal. BP (Magyari et al. 2006, 2009). A similar rise in the water level was deducted for Lake Balaton from 5200 cal.

BP resulting in an expansion of the lake's area exceeding the values of modern day water coverage (Cserny - Nagy-Bodor 2000, Tullner - Cserny 2003, Jakab et. al. 2005). A somewhat delayed similar pattern is observable in the peatlands of the GHP starting at 4400 cal. BP (Jakab et al. 2004a). This period between 5600 and 5300 cal. BP is referred to as the Middle Holocene Climatic Transition, characterizd by a sudden deterioration of the previously warm conditions as a result of the collective transformation of orbital forces, solar activity and ocean currents (Magny et al. 2006, Iizuka et al. 2008).

Three short-lived peat formation events were identified at 8200, 6800 and 3800 cal. BP reflecting cooler conditions. Paleoecological records available from the Carpathian Basin yielded no information of climate change for this period so far. Conversely, data from Western Europe did. There is a marked cooling related to a global cooling event lasting for merely 200 years known as the '8.2 ky event' (Alley et al. 1997, 2005, Bond et al. 1997, Nesje – Dahl 2001). At 6000 cal. BP a high lake level phase of Swiss lakes (Magny 1998, Magny – Schoellammer 1999) and changes in the moisture gradient of some British peatlands (Hughes et al. 2000) refer to the emergence of cooler conditions. Similarly at 3500 cal. BP the higher lake phase of Swiss lakes (Magny 1998, Magny et al. 2002), the expansion of Alpine glaciers (Haas et al. 1998) and an increase in the moisture gradient of numerous Western European peatlands marks a cooling of the climate (Hughes et al. 2000, Barber – Charman 2005).

An increase in the amount of *Sphagna* from 2800 cal. BP in Nyires Peatbog also marks the cooling of the climate and the accompanying rise in the rainfall, that disagrees Feurdean et al.'s (2008) reconstruction. This deterioration of the climate starting at 3500 cal BP culminates here in the Carpathian Basin as was shown by numerous records. Water levels were the highest in the Lake Sf Ana in Romania at this time, and there is information for the development of layering in the waterbody for this period (Magyari et al. 2006, 2009). Congrently with this data, information of studies of testacea and humic content of peatlands in the Eastern Carpathians show an increase in the moisture gradient (Schnitchen et al. 2003). From the Gutin Mts Björkmann et al. (2002a, b) identified an expansion of open peatland areas from 3400 cal. BP parallel with the expansion of beech. The resuming peat formation in certain Hungarian peatlands marks the cooling of the climate here (Jakab – Sümegi 2007). These data on the whole suggest increasing moisture availability in the Carpathians and the adjoining Carpathian basin from *ca.* 3400 years BP with maximum moisture availability around 2700-2800 years BP.

The first real *Sphagnum* peatland developed at Sirok between 2300 and 1500 cal. BP. From here on we have a record of alternating phases of *Sphagnum* peatlands and sedge, reed peatlands. As displayed by the record of vegetation changes, the catchment of the referred peatland was highly prone to climatic fluctuations. Certain periods are characterized by a rapid expansion of *Sphagna,* while others are by the expansion of sedge and reed. A sudden expansion of *Sphagna* was recorded at least 10 times. Fig 6. depicts a comparison of changes inferred for Nyírjes Peatbog with cooler periods determined by Barber et al. (1994) and Mauquoy – Barber (1999), emphasizing changes for the last 3000 years. The *Sphagnum*-peaks perfectly match the more humid periods identified in the British Isles at 2150, 1750, 1300, 1000, 850, 500 and 200 cal. BP (Barber et al. 1994, Barber – Charman 2005, Mauquoy – Barber 1999), referring to some collective global force as the culprit of these changes. Barber and Charman (2005) identified centennial-scale climatic fluctuations in different parts of Western Europe. The length of these cycles was varying ranging between, 210, 600, 800 or 1100 years in different peatlands. No such cycles were identified in Central Europe so far.

It's worth comparing paleoecological data of our referred study site for the last 2000 years with those of written historical records. The interrelatedness of cultural evolution and that of the natural environment is a widely accepted fact. Larger environmental crises led to the collapse of whole empires, famine or wars. Well-known examples are those of the disappearance of the Maya Empire, the famines of the Little Ice Age or the abandonment of Viking settlements in Greenland. This interrelatedness is very often indirect, as seen by the ecological carrying capacity of a region and agricultural production influenced by climatic endowments (Berglund 2003).

One major climatic crisis events in the Carpathian Basin is connected to the fall of the Avar Empire at 8th century AD. Written records blame the famines and wars triggered by the outstanding droughts of the period (Győrffy 1995, Győrffy – Zólyomi 1994). Little environmental historical data was available for this period so far. As shown by the *Sphagnum* curve of our referred study site, this period was truly characterized by dry conditions (Figure 6). The same event was recorded at Nádas Lake of Nagybárkány as well by Jakab et al. (2009) in the area of northern Hungary.

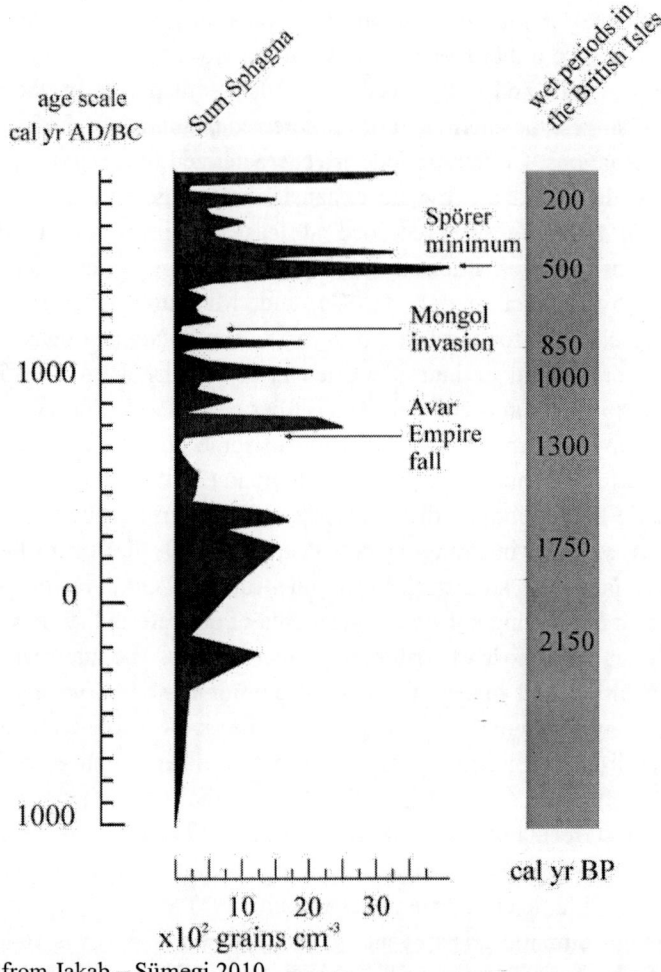

Redrawn from Jakab – Sümegi 2010.

Figure 6. Comparison of bog surface wetness changes of Sirok Nyírjes Peatbog (N Hungary) and some British peatbog (Barber et al. 1994, Mauquoy – Barber 1999) in the last 3000 yr. The arrows show some medieval events.

Another major historical crisis was the advancement of Mongolian tribes to the area in 1241-1242. Ceratin sources blame this on the severe cold weather, while others talk about the outsanding droughts. Data is highly contradictory, as in Europe information is available on the extreme summer droughts for the referred period, while in Hungary the extreme cold winter of 1241 is emphasized, when the complete freezing of the Danube enabled the Mongol herds to cross the river and destroy the settlements of Transdanubia.

This seemingly contradictory information can easily be resolved as shown by Kiss (2000, 2003). The freezing of the Danube in the winter of 1241 was not a unique event, and the summer droughts recorded in Europe must have had similar devastating effects in Hungary, amplifying the negative outcome of the war the famine. As shown by our paleoecological data for the Nyíres Peatbog Hungary was characterized by extreme warm conditions during this period resulting in an almost complete dessication of the *Sphagnum* peatland. Barber et al. (2000) also marked this period as the driest one in the past 3000 years history of Europe.

The *Sphagnum* curve of the Nyíres Peatbog enables us to identify the period of the Little Ice Age as well dated between the middle part of the 16th century till the middle part of the 19th century (Bradley et al. 2003). This was inferred to be the coldest period of the past 2000 years. The coldest period was put to the terminal part of the 16th century (Spörer minimum), when there is a marked drop in the temperature all over Europe (Pfister 1999, Pfister - Brázdil 1999). This event caused severe problems in food production, triggered by transforming sunspot activities. The later cold periods were less devastating on the economies of Europe and Hungary (Rácz 2001). Environmental record for Nyíres Peatbog is congruent with these events as well. The most diverse *Sphagnum* taxa including the hygrophilous *Sphagnum cuspidatum* lived here at the end of the 16th century. Western European peatlands were similarly prone to the fluctuating climate of the Little Ice Age (Mauquoy et al. 2002). Wetter conditions were identified from the beginning of the 16th and middle part of the 17th centuries.

An increase in the surface wetness of our study area of the past 100 years is attributable to not really climatic changes but rather human-indiced soil erosion and alteration of surficial rainfall conditions via deforestation. This is clearly recorded in the intercalating sand grains and clayey horizons of the studied peatland.

It is interesting to ask what climatic factor should be blamed for changes in the moisture gradient of peatlands? Or what extent of temperature increases and accompanying precipitatin decreases can be inferred for the drier periods? According to Blaauw et al. (2004) there is a strong relationship between the moisture gradient of peatlands and solar activity reflected in the correlation of the former parameter with proxy for $\Delta^{14}C$. One may righteously ask what component of the climate controls the moisture gradient of peatlands via fluctuating solar activities? Surficial wetness is controlled by a complex interplay of precipitation and evapotranspiration of the plants, seen in such parameters as annual average rainfall and evaporation, influenced by the temperatures of the growth season. There are no surficial water courses feeding Nyíres Peatbog so runoff must have been influential only during the past 100 years on the hydrology of the peatland.

As was shown in Western Europe the moisture gradient of peatlands for the past 3000 years was primarily determined by fluctuations in the temperature of the vegetation season or and not really the amout of rainfall (Barber et al. 2000, Barber – Langdon 2001, 2007, Barber – Charman 2005, Charman et al. 2009, Schoning et al. 2005, Swindles et al. 2009). According to Charman (2007) in the Atlantic part of Europe summer precipitation and summer temperatures are controlling the moisture gradient of peatlands. Unfortunately macrofossil studies are not acapable to accurately predict former temperatures or precipitation rates. Only the major trajectories of climate changes can be identified. The modern distribution of *Sphagnum* peatlands in Hungary enable us to give a rough estimate. *Sphagnum* peatlands appear in areas characterized by a precipitation of 600 mm per annum on over among modern temperature values. Below this threshold we can come across only sporadic occurrences, while there are no *Sphagna* known below the lower limit of 550 mm. Based on our findings for Nyírjes Peatbog we may infer conditions in the lower hilly areas during the drier periods of the past 3000 years to be similar to that of the central parts of the GHP. The complete disappearance of *Sphagna* from the area must be linked to a steady drop in the rainfall resulting in an least 50 mm deficit in the local water balance. This can be achieved by an increased evapotranspiration as a result of elevated temperatures of the summer growth season. This value must have exceeded even 100 mm during the Middle Holocene Transition.

ACKNOWLEDGMENTS

The first author of this chapter is the grantee of the János Bolyai Scholarship. We thank Sándor Gulyás for the correction of English version.

REFERENCES

Aaby, B. (1976): Cyclic climatic variations in climate over the past 5,500 yrs reflected in raised bogs. – *Nature*, 263: 281–284.

Aaby, B. - Digerfeldt, G. (1986): Sampling techniques for lakes and bogs. - In: Berglund B. E. (ed.) Handbook of Holocene Palaeoecology and Palaeohydrology. John Wiley and Sons Ltd. pp.: 181-194.

Abramova, A. L. – Abramov, I. I. (1959): Kimmeridgian mosses of the Duab River (Abkhazia). – Trudy Bot. Inst. Akad. Nauk SSSR II. 12: 301-359.

Alley, R. B. - Ágústsdóttir, A. M. (2005): The 8k event: cause and consequences of a major Holocene abrupt climate change. - *Quaternary Science Reviews*, 24 (10-11): 1123-1149.

Alley, R. B. - Mayewski, P. A. - Sowers, T. - Stuiver, M. - Taylor, K. C. – Clark, P. U. (1997): Holocene climatic instability; a prominent, widespread event 8200 yr ago. – *Geology*, 25: 483-486.

Bacsó, N. (1959): Magyarország éghajlata. – Akadémiai Kiadó, Budapest, pp. 302.

Balogh, M. (2000a): A lápok rendszerezése. - In: Szurdoki, E. (ed.) Tőzegmohás élőhelyek Magyarországon: kutatás, kezelés, védelem. CEEWEB Munkacsoport, Miskolc. pp.: 57-65.

Balogh, M. (2000b): Az úszólápi szukcesszió kérdései I. – Kitaibelia 5(1): 9-16.

Barber, K. E. (2007): Peatland Records of Holocene Climate Change. - In: Elias, S. A. (ed.): *Encyclopedia of Quatermary Science*, Vol. 3. Elsevier, pp.: 1883-1894.

Barber, K. E. – Chambers, F. M. – Maddy, D. – Brew, J. (1994): A sensitive high resolution record of the Holocene climatic change from a raised bog in northern England. - *The Holocene*. 4: 198-205.

References

Barber, K. E. – Charman, D. (2005): Holocene palaeoclimate records from peatlands. – In: Mackay, A. - Battarbee, R. - Birks J. - Oldfield, F. (eds.) *Global Change in the Holocene. Hodder Arnold*, pp.: 210-226.

Barber, K. E. – Langdon, P. G. (2001): Peat stratigraphy and climate change. – In: Brothwell, D. R. – Pollard, A. M. (eds.) Handbook of Archaeological Sciences, Wiley, Chicester, pp.: 155-166.

Barber, K. E. – Langdon, P. G. (2007): What drives the peat-based palaeoclimate record? A critical test using multi-proxy climate records from nothern Britain. - *Quaternary Science Reviews*, 26: 3318-3327.

Barber, K. E. – Maddy, D. – Rose, N. – Stevenson, A. C. – Stoneman, R. – Thompson, R. (2000): Replicated proxy-climate signals over the last 2000 yr from two distant UK peat bogs: new evidence for regional palaeoclimate teleconnections. – *Quaternary Science Reviews*, 19: 481-487.

Beaudoin, A. B. (2007): On the laboratory procedure for processing unconsolidated sediment samples to concentrate subfossil seed and other plant macroremains. – *Journal of Paleolimnology*, 37: 301-308.

Bennett, K. D. (1992): PSIMPOLL - A quickBasic program that generates PostScript page description of pollen diagrams. - INQUA Commission for the study of the Holocene: working group on data handling methods, *Newsletter*, 8: 11-12.

Berglund, B. E. (2003): Human impact and climate changes – syncronous events and a causal link? – *Quaternary International*, 105: 7-12.

Birks, H. H. (1980): Plant macrofossils in Quaternary lake sediments. - *Arch. Hydrobiol. Beith. Ergebn. Limnol.* 15. pp. 1-60.

Birks, H. H. (2007): Plant macrofossil introduction. - In: Elias, S. A. (ed.): *Encyclopedia of Quatermary Science*, Vol. 3. Elsevier, pp.: 2266-2288.

Birks, H. H. – Mathewes, R. W. (1978): Studies in the vegetational history of Scotland V. Late Devensian and early Flandrian pollen and macrofossil stratigraphy at Abernethy Forest, Inverness-shire. – *New Phytologist*, 80: 455-484.

Birks, H. J. B. (1982): Quaternary Bryophyte Paleo-ecology. - In: Smith A. J. E. (ed.) Bryophyte Ecology, Chapman and Hall, London & New York, pp.: 437-490.

Birks, H. J. B. - Birks, H. H. (1980): Quarternary palaeoecology. - University Park Press, Baltimore, 289 pp.

Birks, H. H. - Birks, H. J. B. (2000): Future uses of pollen analysis must include plant macrofossils. – *Journal of Biogeography*, 27: 31-35.

Björkman, L. – Feurdean, A. – Cinthioa, K. – Wohlfarth, B. – Possnert, G. (2002a): Lateglacial and early Holocene vegetation development in the

Gutaiului Mountains, northwestern Romania. - Quaternary Science Reviews 21: 1039-1059.

Björkman, L. – Feurdean, A. – Wohlfarth, B. (2002b): Late-Glacial and Holocene forest dynamics at Steregoiu in the Gutaiului Mountains, Northwest Romania. - *Review of Palaeobotany and Palynology*, 25(14): 1-33.

Blaauw, M. – van Geel, B. – van der Plicht, J. (2004): Solar forcing of climatic change during the mid-Holocene: indicators from raised bogs in The Netherlands. – *The Holocene*, 14(1): 35-44.

Blytt A. (1876): Essays on the Immigration of Norwegian Flora during Alternating Rainy and Dry Periods. Kristiana Cammermeyer.

Bond, G. – Showers, W. – Cheseby, M. – Lotti, R. – Almasi, P. – deMenocal, P. – Priore, P. – Cullen, H. – Hajdas, I. – Bonani, G. (1997): A pervasive millenial-scale cycle in North Atlantic Holocene and glacial climates. – *Science*, 278: 1257-1266.

Borhidi, A. - Balogh, M. (1970): Die Enstehung von dystrophen Schaukelmooren in einem alkalischen(szik-) See. - *Acta Botanica Hungarica*, 16: 13-31.

Borhidi, A. – Sánta, A. (eds.)(1999): Vörös könyv Magyarország növénytársulásairól 1-2. – Természetbúvár Alapítvány Kiadó, Budapest, 362pp, 404pp.

Boros, Á. (1924): Magyar láptanulmányok IV. – Ungarische Moorstudien. Az egerbektai és a keleméri mohalápok növényzete – Die Flora der Moore von Egerbakta und Kelemér. – *Magyar Botanikai Lapok*, 23: 62–64.

Boros, Á. (1925): Two fossil species of mosses from the diluvial lime tufa of Hungary.- *The Bryologist*, 28(3): 29-32.

Boros, Á. (1937) Megjegyzések a dunántúli Sphagnum–előfordulásokhoz. – *Botanikai Közlemények*, 34: 153–153.

Boros, Á. (1952): Pleisztocén mohák Magyarországon. - *Földtani Közlöny*, 82(7-9): 294-301.

Boros, Á. (1968): Bryogeographie und Bryoflora Ungarns. - *Akadémiai Kiadó, Budapest*, 466pp.

Boros, Á. – Járai-Komlódi, M. – Tóth, Z. – Nillson, S. (1993): An Atlas of Recent European Bryophyte Spores. – Scientia Publishing, Budapest, 321 pp.

Bradley, R. S. – Hughes, M. K. – Diaz, H. F. (2003): Climate in medieval time. – *Science*, 302: 404-405.

Campbell, I. D. – Campbell, C. – Yu, Z. – Vitt, D. H. – Apps, M. J. (2000): Millenial-scale rhytms in peatlands in the western interior of Canada in the global carbon cycle. – *Quaternary Research*, 54:155-158.

Charman, D. J. – Barber, K. E. – Blaauw, M. – Langdon, P. G. - Mauquoy, D. – Daley, T. J. –Hughes P. D.M. – Karofeld, E. (2009): Climate drivers for peatland palaeoclimate records. – *Quaternary Science Reviews*, 28(19-20): 1811-1819.

Charman, D. J. (2007): Summer water deficit controls on peatland water table changes: implications for Holocene palaeoclimate reconstructions. - *The Holocene*, 17(2): 217-227.

Cheddadi, R. - Yu, G. - Guiot, J. – Harrison, S. P. – Prentice, I. C. (1997): The climate of Europe 6000 years ago. – *Climate Dynamics*, 13: 1-9.

Clymo, R. S. (1973): The grows of Sphagnum: some effects of environment. - *Journal of Ecology*, 61: 849-869.

Cushing, E. J. - Wright, H. E. (1965): Hand operated piston corers for lake sediments. *Ecology*, 46: 380-384.

Croll, J. (1875): Climate and time in their geological relations a theory of secular changes of the earth's climate. E. Stanford Press, London.

Cserny, T. – Nagy-Bodor, E. (2000): Limnogeology of Lake Balaton (Hungary). – In: Gierlowski-Kordesch, E. H. – Kelts, K. R. (eds): Lake basins though space and time: *AAPG Studies in Geology*, pp.: 605-618.

Davis, B. A. S. - Brewer, S. - Stevenson, A. C. - Guiot, J. – Data Contributors (2003): The temperature of Europe during the Holocene reconstructed from pollen data. - *Quaternary Science Reviews*, 22: 1701–1716.

Dickson, J. H. (1973): Bryophytes of the Pleistocene. The British record and its chorological and ecological implications. – Cambridge University Press, London & New York, 256 pp.

Dickson, J. H. (1986): Bryophyte analysis. - In: Berglund, B. E. (ed.) Handbook of Holocene Palaeoecology and Palaeohydrology. John Wiley and Sons Ltd. pp.: 627-642.

Dömsödi, J. (1988): Lápképződés, lápmegsemmisülés. – MTA Földrajztudományi Kutató Intézet, Budapest, 120pp.

Feurdean, A. - Klotz, S. - Mosbrugger, V. - Wohlfarth, B. (2008): Pollen-based quantitative reconstructions of Holocene climate variability in NW Romania. - *Palaeogeography, Palaeoclimatology, Palaeoecology*, 260: 494-504.

Gaillard, M-J. - Birks, H. H. (2007): Paleolimnological applications. - In: Elias, S. A. (ed.): *Encyclopedia of Quatermary Science*, Vol. 3. Elsevier, pp.: 2337-2355.

Gams, H. (1932): Quaternary distribution. - In: Verdoorn, F. (ed.) Manual of Bryology, Martinus Nijhoff, The Hague pp. 297-322.

Gardner, A. R. (2002): Neolithic to Copper Age woodland impacts in northeast Hungary? Evidence from the pollen and sediment chemistry records. – *The Holocene*, 12: 521-553.

Gignac, D. L. – Vitt, D. H. (1990): Habitat limitations of Sphagnum along climatic, chemical and physical gradients in mires of western Canada. – *The Bryologist*, 93(1): 7-22.

Grosse-Brauckmann, G. (1972): Über pflanzliche Makrofossilien mitteleuropäischer Torfe. I: Gewebereste krautiger Pfanzen und ihre Merkmale. – *Telma*, 2: 19-55.

Grosse-Brauckmann, G. (1986): Analysis of vegetative plant macrofossils.- In: Berglund, B. E. (ed.) Handbook of Holocene Palaeoecology and Palaeohydrology. John Wiley and Sons Ltd. pp.: 591-618.

Grosse-Brauckmann, G. – Haussner, W. – Mohr, K. (1973): Über eine kleine Vermoorung im Odenwald, ihre Ablagerungen und ihre Entwicklung der umgebenden Kulturlandschaft. – *Z. Kulturtechnik und Flurbereinigung*, 14: 132-143.

Győrffy, Gy. (1995): Hová lettek az avarok? – *História*, 17(3): 3-9.

Győrffy, Gy. – Zólyomi, B. (1994): A Kárpát-medence és Etelköz képe egy évezred előtt. – In: Győrffy, Gy. – Kovács, L. (eds.): Honfoglalás és régészet. Balassi Kiadó, Budapest. pp. 13-37.

Haas, J. N. – Richoz, I. – Tinner, W. – Wick, L. (1998): Synchronous Holocene climatic oscillations recorded on the Swiss Plateau and at the timberline in the Alps. – *The Holocene*, 8: 301-304.

Hájková, P. – Hájek, M. (2004): Bryophyte and vascular plant responses to base-richness and water level gradients in western Carpathian Sphagnum-rich mires. – *Folia Geobotanica*, 39: 335-351.

Hakan, R. (1993): Interspecific competition between Sphagnum mosses on a raised bog. – *Oikos*, 66: 413-423.

Hall, A. R. (1979): A note on the Quaternary history of Meesia longiseta Hedw. in Britain. - *Journal of Bryology*, 10: 511-515.

Halsey, L. A. - Vitt, D. H. - Bauer, I. E. (1998): Peatland initiation during the Holocene in continental western Canada. - Climatic Change 40: 315–342.

Haslam, S. M. (1972): Biological flora of the British Isles. Phragmites communis Trin. - *Journal of Ecology*, 60: 585–610.

Herzog, Th. (1926): Geographie der Moose. Fischer Jena 440pp.

Hughes, P. D. M. – Mauquoy, D. – Barber, K. E. – Langdon, P. G. (2000): Mire development pathways and palaeoclimatic records from a full Holocene peat archive at Walton Moss, Cumbria, England. – *The Holocene*, 10: 465-479.

Iizuka, Y. - Hondoh, T. - Fujii, Y. (2008): Antarctic sea ice extent during the Holocene reconstructed from inland ice core evidence. – Journal of Geophysical Research 113. D15114, doi:10.1029/2007JD009326.

References

Jakab, G. – Magyari, E. 2000. Új távlatok a magyar lápkutatásban: szukcessziókutatás paleobryológiai és pollenanalitikai módszerekkel.– *Kitaibelia*, 5(1): 17-36.

Jakab, G. – Sümegi, P. – Magyari, E. (2004a): A new paleobotanical method for the description of Late Quaternary organic sediments (Mire-development pathways and palaeoclimatic records from S Hungary). - *Acta Geologica Hungarica*, 47(4): 1-37.

Jakab, G. – Sümegi, P. (2004b): A lágyszárú növények tőzegben található maradványainak határozója mikroszkópikus bélyegek alapján. – *Kitaibelia*, 9(1): 93-129.

Jakab, G. – Sümegi, P. (2005): The evolution of Nádas-tó at Nagybárkány in the light of the macrofossil finds. - In: Gál, E. – Juhász, I. E. - Sümegi, P. (eds.) Environmental History in North-Eastern Hungary, *Varia Archaeologica Hungarica*, 19: 67-77

Jakab, G. – Sümegi, P. (2007): The vegetation history of Baláta-tó. - In: Juhász, I. E. – Zatykó, Cs. – Sümegi, P. (eds.) Environmental History of Transdanubia, *Varia Archaeologica Hungarica*, 20: 251-254.

Jakab, G. – Sümegi, P. (2010): The Role of Bryophyte Paleoecology in Quaternary Climate Reconstructions. – In: Tuba, Z. - Slack, N. G. (eds.) Bryophyte Ecology Under Changing Climate. Cambridge University Press, London & New York, (in press)

Jakab, G. – Sümegi, P. – Szántó, Zs. (2005): Késő-glaciális és holocén vízszintingadozások a Szigligeti-öbölben (Balaton) makrofosszília vizsgálatok eredményei alapján. – *Földtani Közlöny*, 135(3): 405-431.

Jakab, G. – Majkut, P. - Juhász, I. - Gulyás, S. - Sümegi, P. - Törőcsik, T. (2009): Palaeoclimatic signals and anthropogenic disturbances from the peatbog at Nagybárkány (N Hungary). – *Hydrobiologia*, 631(1): 87-106.

Janssens, J. A. P. (1977): Bryophytes from the Pleistocene of Belgium and France. – *Journal of Bryology*, 9: 349-359.

Janssens, J. A. P. (1983a): A quantitative method for stratigraphic analysis of bryopytes in holocene peat. - *Journal of Ecology*, 71: 198-196.

Janssens, J. A. P. (1983b): Past and extant distribution of Drepanocladus in North America, with notes on the differentation of fossil fragments. - *Journal of the Hattori Botanical Laboratory*, 54: 251-298.

Janssens, J. A. P. (1987): Ecology of peatland bryophytes and palaeoenvironmental reconstruction of peatlands using fossil bryophytes. - Manual for Bryological Methods Workshop. Satellite Conference of the XIV. Intenational Botanical Conference International Association of Bryologists, Mainz, 67 pp.

Janssens, J. A. P. (1990): Methods in Quarternary Ecology 11. Bryophytes. - Geoscience Canada 17(1): 13-24.
Jasnowski, M. (1957): Moosflora quartärer Flachmoorablagerungen. – *Acta Soc. Bot. Pol.* 26: 597-629.
Jessen, K. (1949): Studies in the Late Quaternary deposits and flora-history of Ireland. – *Proc. Royal Irish Acad.* 52(B): 85-290.
Jessen, K. – Milthers, V. (1928): Statigraphical and paleontological studies of interglacial fresh-water deposits in Jutland and northwest Germany. – Danmarks Geol. Unders., Series 11. 48: 1-378.
Jowsey, P. C. (1966): An improved peat sampler. - New Phytologist 65: 245-248.
Joerin, U. E. - Nicolussi, K. - Fischer, A. - Stocker, T. F. - Schlüchter, C. (2008): Holocene optimum events inferred from subglacial sediments at Tschierva Glacier, Eastern Swiss Alps. - *Quaternary Science Reviews*, 27: 337-350.
Katz, N. J. (1930): Über die Typen der Moore der westsibirischen Niederung und ihre geografische Zonation. – *Ber. Deutsch. Bot. Ges.* 48: 13-25.
Keller, B. - Bürgi, H. R. - Sturm, M. - Spaak, P. (2002): Ephippia and Daphnia abundances under changing trophic conditions. - Verhandlungen Internationale Vereinigung für Theoretische und Angewandte Limnologie 28: 851-855.
Kiss, A. (2000): Weather events during the first Tartar invasion in Hungary (1241-42). – *Acta Geographica Szegediensis*, 37: 149-156.
Kiss, A. (2003): „Ecce, in hyemis nivis et glaciei habundantia supervenit" – Időjárás, környezeti krízis és a tatárjárás. - In: Nagy, B. (ed.) Tatárjárás. Osiris Kiadó, Budapest, pp.: 439-452.
Kooijman, A. M. (1993): On the ecological amplitude of four mire bryophytes; a reciprocal transplant experiment. – *Lindbergia*, 18: 19-24.
Korhola, A. (1995): Holocene climatic variations in southern Finland reconstructed from peat-initiation data. – *The Holocene*, 5(1): 43-57.
Kutzbach, J. E. - Ruddiman, W. F. – Prell, W. L. (1997): Possible effects of Cenozoic uplift and CO_2 lowering on global and regional hydrology. – In: Ruddiman, W. F. (ed.) Tectonic Uplift and Climate Change. Plenum Publishing Corporation, New York, pp.: 149-170.
Lájer, K. (1998a): Az Aldrovanda vesiculosa L. újabb előfordulása és egyéb adatok Magyarország flórájának ismeretéhez. – Kitaibelia 3: 263-274.
Lájer, K. (1998b) Bevezetés a magyarországi lápok vegetáció-ökológiájába. – Tilia 6: 84-238.
Libby, W. F. (1946): Atmospheric Helium Three and Radiocarbon from Cosmic Radiation. - *Physical. Review*, 69: 671-672.

Lowe, J. J. - Walker, M. J. C. (1997): Reconstructing Quaternary environments 2. - Longmann Press, London.

Magny, M. (1998): Reconstruction of Holocene lake-level changes in the Jura (France): methods and results. – In: Harrison, S. P. – Frenzel, B. – Huckried, U. – Weiss, M. (eds.) Palaeohydrology as Reflected in Lake-level Changes as Climatic Evidence for Holocene Times, Paläoklimaforschung, 25: 67-85.

Magny, M. - Schoellammer, P. (1999): Lake-level fluctuations at Le Locle, Swiss Jura, from the Younger Dryas to the Mid-Holocene: A high-resolution record of climate oscillations during the final deglaciation. - *Géographie Physique et Quaternaire*, 53(2): 183-197.

Magny, M. – Miramont, C. – Sivan, O. (2002): Assesment of climate and antropogenic factors on Holocene Mediterranean vegetation in Europe on the basis of palaeohydrological records. – *Palaeogeography, Palaeoclimatology, Palaeoecology*, 186: 47-59.

Magny, M. - Begeot, C. - Guiot, J. - Peyron, O. (2003): Constraining patterns of hydrological changes in Europe in response to Holocene climate cooling phases. - *Quaternary Science Reviews*, 22: 1589–1596.

Magny, M. - Leuzinger, U. - Bortenschlager, S. - Haas, J. N. (2006): Tripartite climate reversal in Central Europe 5600–5300 years ago. – *Quaternary Research*, 65: 3-19.

Magyari, E. - Jakab, G. - Sümegi, P. - Rudner, E. - Molnár, M. (2000): Paleobotanikai vizsgálatok a keleméri Mohos-tavakon. - In: Szurdoki, E. (ed.) Tőzegmohás élőhelyek Magyarországon: kutatás, kezelés, védelem. CEEWEB Munkacsoport, Miskolc. pp.101-131.

Magyari, E. – Sümegi, P. – Braun, M. – Jakab, G. – Molnár, M. (2001): Retarded wetland succession: antropogenic and climatic signals in a Holocene peat bog profile from north-east Hungary. - *Journal of Ecology*, 89: 1019-1032.

Magyari, E. - Buczkó, K. - Jakab, G. - Braun, M. - Szántó, Zs. - Molnár, M - Pál, Z. - Karátson, D. (2006): Holocene palaeohydrology and environmental history in the South Harghita Mountains, Romania. - *Földtani Közlöny*, 136(2): 249-284.

Magyari, E. K. - Buczkó, K. - Jakab, G. - Braun, M. - Pál, Z. - Karátson, D. (2009). Palaeolimnology of the last Eastern Carpathian crater lake - a multiproxy study of Holocene hydrological changes. – *Hydrobiologia*, 631(1): 29-63.

Mauquoy, D. – Barber, K. (1999): A replicated 3000 yr proxy-climate record from Coom Rigg Moss and Felicia Moss, The Border Mires, northern England. – *Journal of Quaternary Science*, 14(3): 263-275.

Mauquoy, D. – van Geel, B. (2007): Mire and peat macros. - In: Elias, S. A. (ed.): *Encyclopedia of Quatermary Science*, Vol. 3. Elsevier, pp.: 2315-2336.
Mauquoy, D. – van Geel, B. – Blaauw, M. – van der Plicht, J. (2002): Evidence from northwest European bogs shows 'Little Ice Age' climatic changes driven by variations in solar activity. – *The Holocene*, 12(1): 1-6.
Máthé, I. – Kovács, M. (1958): A Mátra tőzegmohás lápja. – *Botanikai Közlemények*, 47(3-4): 323-331.
Máthé, I. - Kovács, M. (1959): A Cserhát tőzegmohás lápja. - *Botanikai Közlemények*, 50: 106-108.
Milankovitch, M. (1930): Mathematische Klimalehre und Astronomische Theorie der Klimaschwankungen, Handbuch der Klimalogie Band 1. Teil A Borntrager Berlin.
Miller, N. G. (1976): Studies of North American Quaternary bryophyte subfossils I. A new moss assemblage from the Two Creeks Forest Bed of Wisconsin. – Occas. Papers Farlow Herb. Harvard Univ. 9: 21-42.
Miller, N. G. (1983): Tertiary and Quaternary Fossils. - In: Schuster, R. M. (ed.) New Manual of Bryology. The Hattori Botanical Laboratory, Nichinan, pp.: 1194-1232.
Mitsch, W. J. – Gosselink, J. G. (1993): Wetlands. – Van Nostrand Reinhold, New York, 722pp.
Moore, J. J. (1968): A classification of the bogs and wet heaths of Northern Europe (Oxycocco-Sphagnetea Br-Bl. ex Tx. 1943). In: Tüxen, R. (ed.) Pflanzensoziologische Systematik. Den Haag, pp.: 306-320.
Moore, P. D. – Bellamy, D. J. (1974): Peatlands. – Elek Science, London. 221pp.
Nesje, A. - Dahl, S. O. (2001): The Greenland 8200 cal. yr BP event detected in loss-on-ignition profiles in Norwegian lacustrine sediment sequences. - *Journal of Quaternary Science*, 16 (2): 155–166.
Nurminen, L. (2003): Macrophyte species composition reflecting water quality in adjecent water bodies of lake Hiidenvesi, SW Finland. - *Annales Botanici Fennici*, 40: 199-208.
Odgaard, B. V. (1980): Ecology, distribution and late Quaternary history of Polytrichastrum alpinum (Hedw.) G. L. Smith in Denmark. – *Lindbergia*, 6: 155-158.
Odgaard, B. V. (1981): The Quaternary bryoflora of Denmark. I. Species list. - Danm. Geol. Undersőg. Áborg 1980: 45-74.
Odgaard, B.V. (1988): Glacial relicts - and the moss Meesia triquetra in Central and West Europe. – *Lindbergia*, 14: 73-78.
Oldfield, F. (2005): Environmental Change: Key Issues and Alternative Approaches. – Cambridge University Press, Cambridge, 363 pp.

Osvald, H. (1925): Die Hochmoortypen Europas. - Veröff. Geobot. Inst. Rubel, Zürich 3: 707-723.
Ódor, P. - Szurdoki, E. - Tóth, Z. (1996): Újabb adatok a Vendvidék mohaflórájához. - *Botanikai Közlemények*, 83: 97–108.
Pakarinen, P. (1995): Classification of boreal mires in Finland and Scandinavia: A review. – *Vegetatio*, 118: 29-38.
Pfister, C. (1999): Wetternachhersage: 500 Jahre Klimvariationen und Naturkatastrophen (1496-1995). - Haupt: Bern. 304pp.
Pfister, C. – Brázdil, R. (1999): Climatic Variability in Sixteenth-Century Europe and its Social Dimension: A Synthesis. - Climatic Change 43(1): 5-53.
Pilous, Z. (1968): Pleistocäne und Holozäne Moose im Gebiete von Ostrava (Ostravsko). – *Preslia*, 40: 68-75.
Podani, J. (1993): SYN-TAX 5.0: Computer programs for multivariate data analysis in ecology and systematics. - Abstracta Botanica 17: 289-302.
Rácz, L. (2001): Magyarország éghajlattörténete az újkor idején. – JGYF Kiadó, Szeged, 303pp.
Rybniček, K. (1966): Glacial relics in the Bryoflora of the Highlands Českomoravská vrchovina (Bohemian – Moravian Highlands: Their habitat and cenotaxonomic value.) – *Folia Geobotanica et Phytotaxonomica*, 1: 101-119.
Rybníček, K. (1973): A comparison of the present and past mire communities of Central Europe. - In: Birks, H. J. B.- West, R. G. (eds.) Quaternary Plant Ecology. Blackwell, Oxford. pp.: 237-261.
Rybníček, K. - Rybníčková, E. (1974): The origin and development of waterlogged meadows in the central part of the Sumava Foothills. - *Folia Geobotanica et Phytotaxonomica*, 9: 45-70.
Rybníčková, E. (1974): Die Entwicklung der Vegetation und Flora im Südlichen Teil der Böhmisch-Mährischen Höhe während des Spätglacial und Holozäns. – *Vegetace CSSR*, 7: 1-163.
Schoning, K. - Charman, D. J. - Wastegard, S. (2005): Reconstructed water tables from two ombrotrophic mires in eastern central Sweden compared with instrumental meteorological data. - *The Holocene*, 15: 111–118.
Simon, T. (1953): Torfmoore im Norden des Ungarischen Tietlandes. – *Acta Biologica Hungarica*, 4: 249-252.
Sernander, R. (1908): On the evidence of postglacial changes of climate furnished by the peat-mosses of northern Europe. – *Geol. Fören. Stockh. Förh.* 30: 467-478.
Slack, N. G. (1994): Can one tell the mire type from the bryophytes alone? – *Journal of the Hattori Botanical Laboratory*, 75: 149-159.

Smith, L. C. - MacDonald, G. M. - Frey, K. E. - Velichko, A. A. - Kremenetski, K. V. - Borisova, O. K. - Dubinin, P. - Forster, R. (2000): U.S.-Russian venture probes Siberian peatlands sensitivity to climate. EOS 81: 497-504.

Smith, L. C. – MacDonald, G. M. - Velichko, A. A. – Beilman, D. W. - Borisova, O. K. - Frey, K. E. - Kremenetski, K. V. – Sheng, Y. (2004): Siberian peatlands a net carbon sink and global methane source since the early Holocene. – *Science*, 303: 353-356.

Schnitchen, C. – Magyari, E. – Tóthmérész, B. – Grigorszky, I. – Braun, M. (2003): Micropaleontological observations on a Sphagnum bog in East Carpathian region - testate amoebae (Rhizopoda: Testacea) and their potential use for reconstruction of micro- and macroclimatic changes. – *Hydrobiologia*, 506(1-3) 45-49.

Steffen, H. (1931): Vegetationskunde von Ostpreußen. – *Gustav Fischer Verlag, Jena*, 406pp.

Stefureac, T. I. (1962): Relictes subarctique dans la bryoflore du marais eutrophe de Dragoiasa Carpathes orientales. – *Rev. Bryol. Lichen,*. 31: 68-73.

Swindles, G. T. – Blundell, A. – Roe, H. M. – Hall, V. A. (2009): A 4500-year proxy climate record from peatlands in the Noth of Ireland: the identification of widespread summer 'drought phases'? - *Quaternary Science Reviews* (in press)

Szafran, B. (1948): Relicts of past epochs in the moss flora of Poland and adjacent eastern regions. – *Ochr. Przyr.* 18: 41-65.

Szafran, B. (1952): Pleistocene moosses from Poland and the adjacent eastern territories. – *Panstw. Inst. Geol.* 68: 5-38.

Szurdoki, E. (2003): Peatmosses of North Hungary. - *Studia Botanica Hungarica*, 34: 55-79.

Szurdoki, E. (2005): Magyarországi tőzegmohafajok elterjedése és egyes fajok vízkémiai igényének vizsgálata. [Distribution of Hungarian peat mosses and investigation of water chemical relation of some species] PhD thesis, L. Eötvös University, Budapest (mscr.)

Szurdoki, E. - Nagy, J. (2002): Sphagnum dominated mires and Sphganum occurrences of North-Hungary. – *Folia Historico Naturalia Musei Matrensis*, 26: 67–84.

Szurdoki, E. - Ódor, P. (2004): Distribution and expansion of Sphagnum fimbriatum Wils. in Hungary. – *Lindbergia*, 29: 136-142.

Szurdoki, E. - Tóth, Z. - Pelles, G. (2000): The Sphagnum populations of the Zemplén Mountains, NE Hungary. – *Studia Botanica Hungarica*, 30-31: 113-125.

Troels-Smith, J. (1955): Karakterisering af lose jordater. - *Danmarks Geologiske Undersogelse*, 4(3): 1-73.

Toivonen, H. - Bäck, S. (1989): Changes in aquatic vegetation of a small eutrophicated and lowered lake (southern Finland). - *Annales Botanici Fennici*, 26: 27–38.

Tullner, T. – Cserny, T. (2003): New aspects of lake-level changes: Lake-Balaton, Hungary. - *Acta Geologica Hungarica*, 46(2): 215-238.

Vaday, A. (2005): The Nagybárkány area in the Iron Age and the Roman period and Migration period. - In: Gál, E., I. Juhász - Sümegi, P. (eds): Environmental archaeology in North-Eastern Hungary. Varia Archaeologica Hungarica XIX. Budapest. pp. 345-347.

Visnya, A. (1939): Sphagnum-folt kalaposkőn. – Vasi Szemle 6: 346-347.

Vitt, H. D.- Chee, W.-L. (1990): The relationships of vegetation to surface water chemistry and peat chemistry in fens of Alberta, Canada. – *Vegetatio*, 89: 87-106.

Warncke, E. (1980): Spring areas: ecology, vegetation and comments on similarity coefficients applied to plant communities. – *Holarct. Ecol.* 3: 233-308.

Wasylikowa, K. (1996): Analysis of fossil fruits and seeds. - In: Berglund B. E. (ed.) Handbook of Holocene Palaeoecology and Palaeohydrology. John Wiley and Sons Ltd. pp.: 571-590.

Yu, Z. – Vitt, D. H. – Campbell, I. D. – Apps, M. J. (2003a): Understanding Holocene peat accumulation pattern of continental fens in western Canada. – *Can. J. Bot.* 81: 267-282.

Yu, Z. - Campbell, I. D. - Campbell, C. - Vitt, D. H. – Bond, G. C. - Apps, M. J. (2003b): Carbon sequestration in western Canadian peat highly sensitive to Holocene wet-dry climate cycles at millennial timescales. – *The Holocene*, 13(6): 801-808.

Zoltai, S. C. – Vitt, D. H. (1990): Holocene climatic change and the distribution of peatlands in western interior Canada. – Quaternary Research 33: 231-240.

Zoltai, S. C. – Vitt, D. H. (1995): Canadian wetlands: Environmental gradients and classification. – *Vegetatio*, 118: 131-137.

Zólyomi, B. (1931): A Bükkhegység környékének Sphagnum lápjai. (Vegetationsstudien an den Sphagnummooren um das Bükkgebirge in Mittel–Ungarn.) – *Botanikai Közlemények*, 28: 89-121.

Zólyomi, B. (1939): Das Kőszeger Sphagnumreiche Moor. – *Botanikai Közlemények*, 36: 318–325.

INDEX

A

Arcto-Alpine elements, 7

B

blanket bogs, 11
bog, vii, 27, 28, 31, 33, 34, 38, 43, 47, 50, 53
brown-moss, 7

C

Carpathian Basin, 7, 19, 33, 34, 35, 36, 37

D

distribution of *Sphagnum*, 40

F

floating mat, 7, 25, 31

G

glacial, 1, 2, 7, 45

H

humic acid, 13
hydroseries, 10, 31

I

ice age calendar, 1
interglacials, 2

L

Little Ice Age, 32, 37, 39, 51

M

Medieval Warm Period, 31, 34

Mid Holocene, 34
minerotrophic peatlands, 11

N

Neoglaciation, 34

O

ombrotrophic peatlands, 11

P

paludification, 11
peat components, 16, 23, 28
peat initiation, 11, 12, 24
plant macrofossils, vii, 5, 13, 44, 47

Q

QLCMA, 15, 16
Quaternary, i, iii, iv, v, vii, 1, 2, 3, 5, 6, 15, 33, 43, 44, 45, 46, 47, 48, 49, 50, 51, 52, 53, 54

R

radiocarbon analysis, 2
raised bogs, 11, 43, 45
relicts, 7, 51
rich fen, 7

S

sampling, 10, 15
spores, 5, 7
synchronous global events, 3

T

terrestrialization, 11